STUDENT LECTURE ART NOTEBOOK

to accompany

Earth

Portrait of a Planet
Second Edition

STEPHEN MARSHAK

University of Illinois

W. W. NORTON & COMPANY

NEW YORK   LONDON

W. W. Norton & Company has been independent since its founding in 1923, when William Warder Norton and Mary D. Herter Norton first published lectures delivered at the People's Institute, the adult education division of New York City's Cooper Union. The Nortons soon expanded their program beyond the Institute, publishing books by celebrated academics from America and abroad. By mid-century, the two major pillars of Norton's publishing program—trade books and college texts—were firmly established. In the 1950s, the Norton family transferred control of the company to its employees, and today—with a staff of four hundred and a comparable number of trade, college, and professional titles published each year—W. W. Norton & Company stands as the largest and oldest publishing house owned wholly by its employees.

Composition by TSI Graphics and Carde Deshoes
Manufacturing by Courier
Illustrations for the Second Edition by Precision Graphics

Editor: Jack Repcheck
Project editor: Thomas Foley
Copy editor: Alice Vigliani
Electronic media editor: April Lange
Director of manufacturing: Diane O'Connor
Photography editors: Neil Ryder Hoos and Stephanie Romeo
Editorial assistant: Sarah Solomon
Book designer: Joan Greenfield
Developmental editor for the First Edition: Susan Gaustad

ISBN 0-393-92781-4

W. W. Norton & Company, Inc., 500 Fifth Avenue, New York, N.Y.
10110
www.wwnorton.com

W. W. Norton & Company Ltd., Castle House, 75/76 Wells Street,
London W1T 3QT

1 2 3 4 5 6 7 8 9 0

CONTENTS

The purpose of this Student Lecture Art Notebook is to provide you with a handy resource for taking notes and mastering the concepts of your introductory geology class.

The key to succeeding in this course is to visualize the geologic processes presented, which is why we have worked so hard to create realistic and instructive diagrams. All of the drawn images from Stephen Marshak's *Earth: Portrait of a Planet, Second Edition* are reproduced in this book, so that you can concentrate on absorbing what they mean, instead of having to spend time redrawing them. Space has been provided for your notes and annotations.

The pages are perforated and three-hole punched for easy transfer and integration with other course material.

PRELUDE   *And Just What Is Geology?*

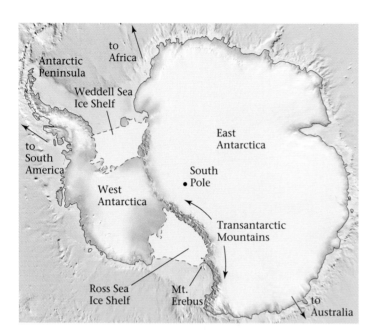

FIGURE P.1 Map of Antarctica.

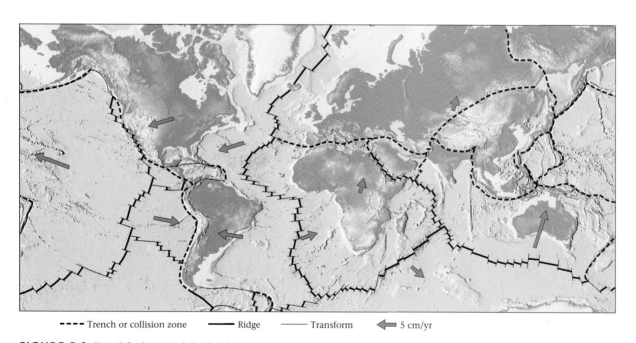

- - - - Trench or collision zone ——— Ridge ——— Transform ⬅ 5 cm/yr

FIGURE P.6 Simplified map of the Earth's principal plates.

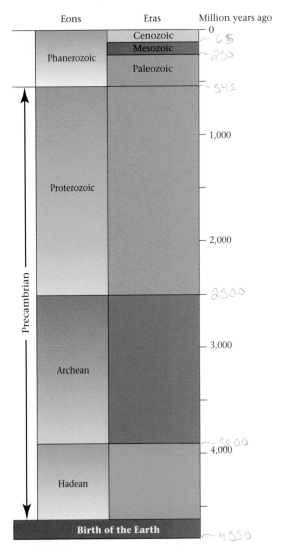

FIGURE P.7 The major divisions of the geologic time scale.

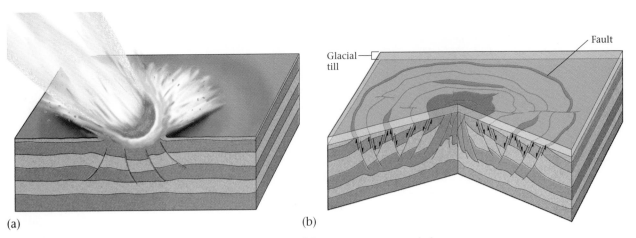

(a)

(b)

FIGURE P.8 (a) The site of an ancient meteorite impact. A large crater, surrounded by debris, forms. (b) The site of the impact today. The crater and the surface debris have eroded away.

CHAPTER 1 *Cosmology and the Birth of Earth*

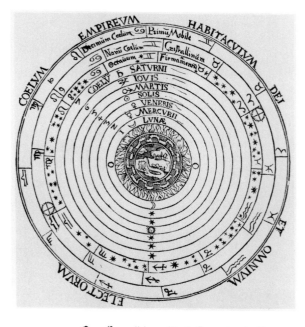

(a)

(b)

FIGURE 1.1 (a) The geocentric image of the Universe. (b) The heliocentric view of the Universe.

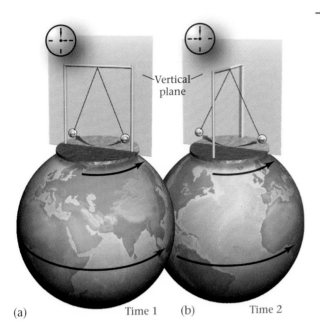

(a) Time 1   (b)   Time 2

FIGURE 1.3 Foucault's experiment. (a) An oscillating pendulum at a given time. (b) The same pendulum at a later time.

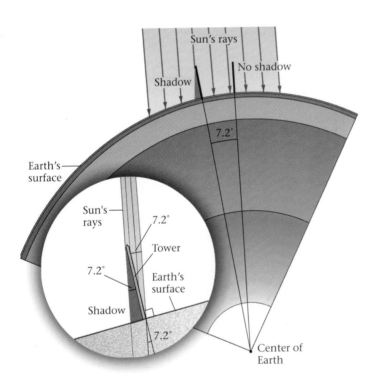

FIGURE 1.4 How Eratosthenes calculated the circumference of the Earth.

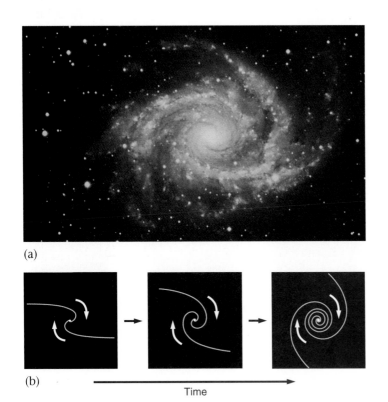

(a)

(b)

Time

FIGURE 1.5 (a) What the Milky Way galaxy might look like if viewed from outside. (b) The spiral shape is due to rotation of the galaxy.

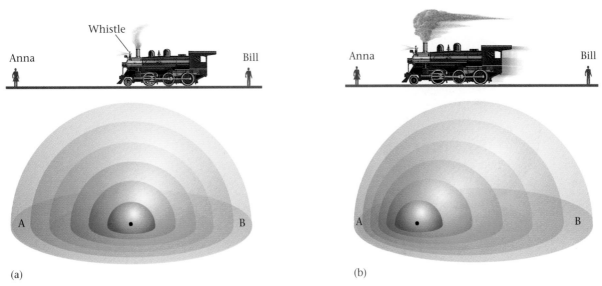

Anna   Whistle   Bill

A   B

(a)

Anna   Bill

A   B

(b)

FIGURE 1.6 The spacing of waves is wavelength. (a) Sound emanating from a stationary source. (b) Sound emanating from a moving source.

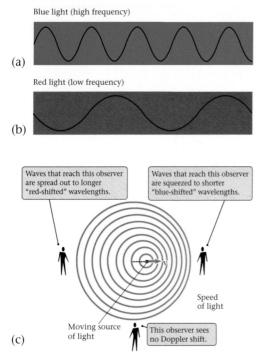

Blue light (high frequency)

(a)

Red light (low frequency)

(b)

Waves that reach this observer are spread out to longer "red-shifted" wavelengths.

Waves that reach this observer are squeezed to shorter "blue-shifted" wavelengths.

Speed of light

Moving source of light

This observer sees no Doppler shift.

(c)

FIGURE 1.7 Light waves. (a) Blue light—short wavelength (higher frequency). (b) Red light—long wavelength (lower frequency). (c) The shift in light frequency that an observer sees.

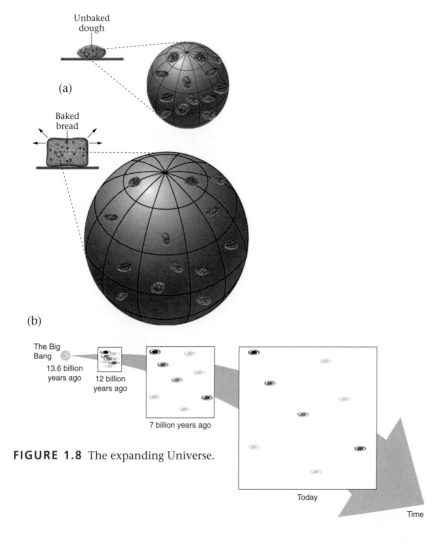

Unbaked dough

(a)

Baked bread

(b)

The Big Bang

13.6 billion years ago

12 billion years ago

7 billion years ago

Today

Time

FIGURE 1.8 The expanding Universe.

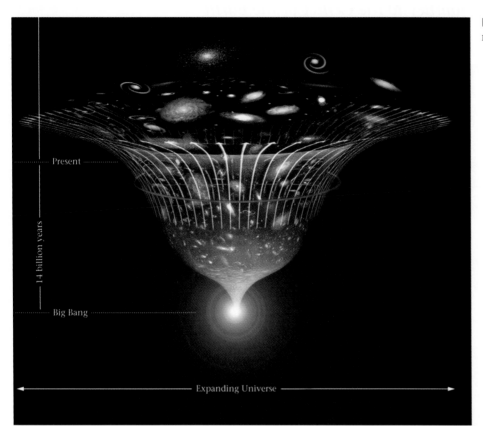

FIGURE 1.9 An artist's rendition of the Big Bang.

FIGURE 1.13 The relative sizes of the planets of our solar system.

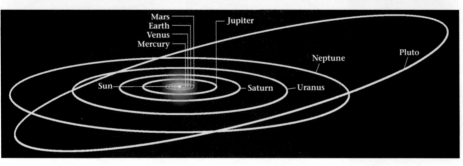

FIGURE 1.15 The planets' ecliptics.

NOTES

CHAPTER 2 *Journey to the Center of the Earth*

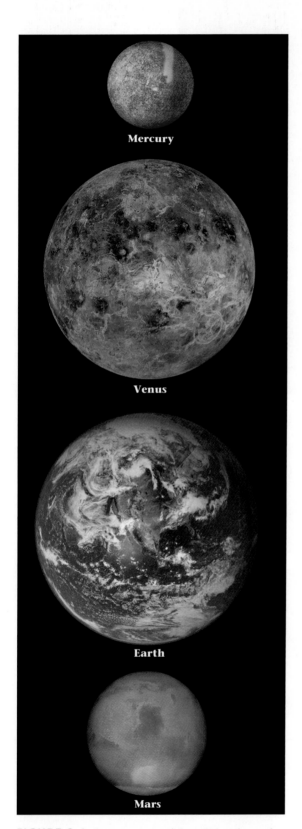

FIGURE 2.1 A comparison of the solid surfaces of Mercury, Venus, Earth, and Mars.

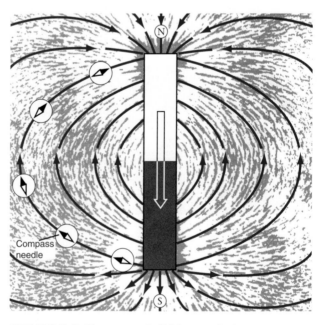

FIGURE 2.2 The magnetic field around a bar magnet.

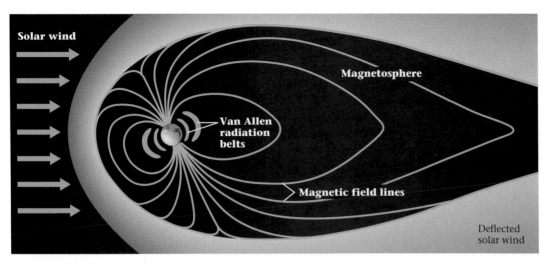

FIGURE 2.3 The magnetic field of the Earth interacts with the solar wind.

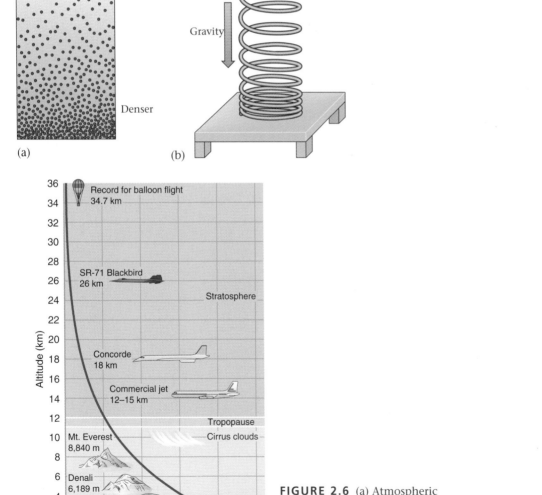

FIGURE 2.6 (a) Atmospheric density increases toward the base of the atmosphere. (b) A spring serves as an analogy. (c) A graph displaying the variation of air pressure with elevation.

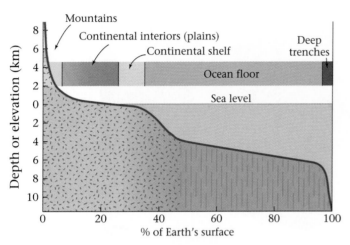

FIGURE 2.9 A hypsometric diagram shows the proportions of the Earth's solid surface at different elevations.

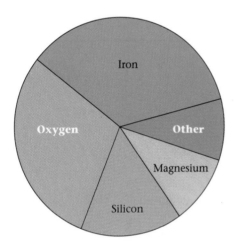

FIGURE 2.10 The proportions of major elements making up the mass of the whole Earth.

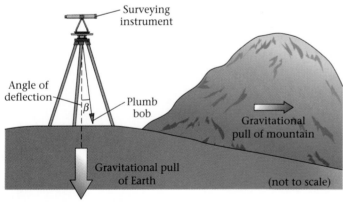

FIGURE 2.12 A surveyor's angled plumb line.

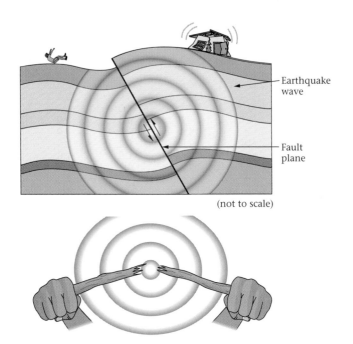

FIGURE 2.13 A fault generates shock waves, much like the sound waves from a stick snapping.

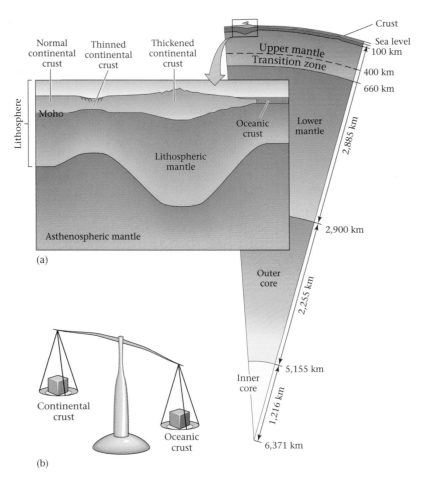

FIGURE 2.14 (a) The differences between continental crust and oceanic crust. (b) Oceanic crust is denser than continental crust.

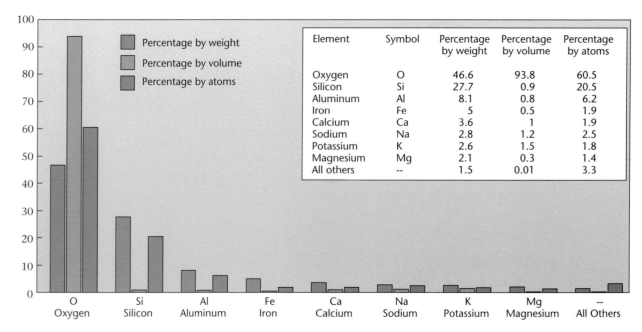

Element	Symbol	Percentage by weight	Percentage by volume	Percentage by atoms
Oxygen	O	46.6	93.8	60.5
Silicon	Si	27.7	0.9	20.5
Aluminum	Al	8.1	0.8	6.2
Iron	Fe	5	0.5	1.9
Calcium	Ca	3.6	1	1.9
Sodium	Na	2.8	1.2	2.5
Potassium	K	2.6	1.5	1.8
Magnesium	Mg	2.1	0.3	1.4
All others	--	1.5	0.01	3.3

FIGURE 2.15 A table and graph illustrating the abundance of elements in the Earth's crust.

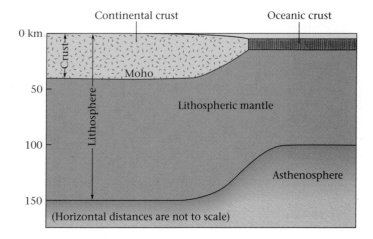

FIGURE 2.16 A cross section of the lithosphere.

NOTES

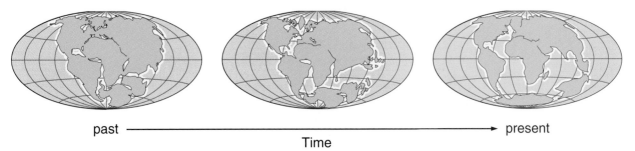

past ——————————————————————→ present

Time

FIGURE 3.2 Wegener's image of Pangaea.

NOTES

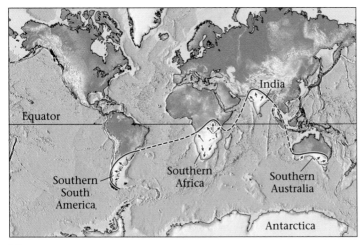

(a)

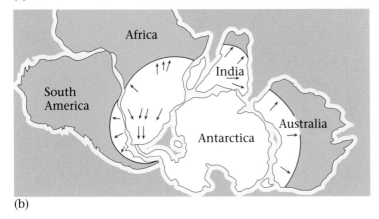

(b)

FIGURE 3.3 (a) The distribution of late Paleozoic glacial deposits. (b) The distribution of these glacial deposits on a map of the southern portion of Pangaea.

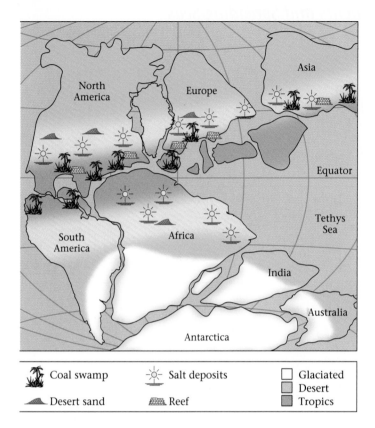

FIGURE 3.4 Map of Pangaea.

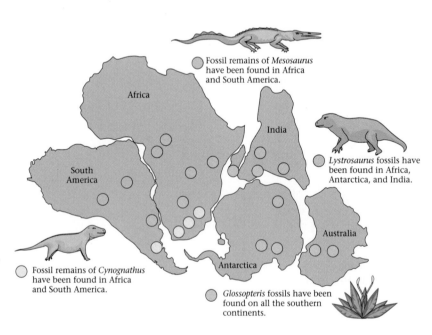

FIGURE 3.5 This map shows the distribution of terrestrial (land-based) fossil species.

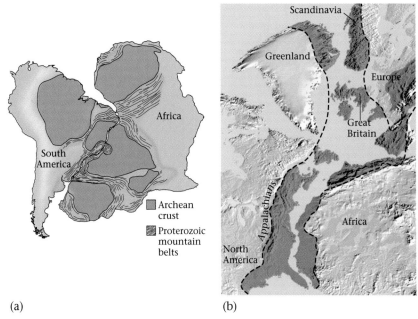

(a)　　　　　(b)

FIGURE 3.6 (a) Distinctive areas of rock assemblages on South America link with those on Africa. (b) If the continents were returned to their positions in Pangaea by closing the Atlantic.

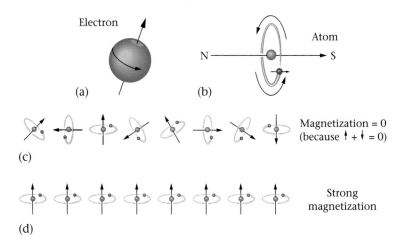

FIGURE 3.7 (a) A spinning electron creates an electric current. (b) The magnetic dipole of an atom. (c) In nonmagnetic material, atoms tilt all different ways. (d) In a permanent magnet, the dipoles lock into alignment.

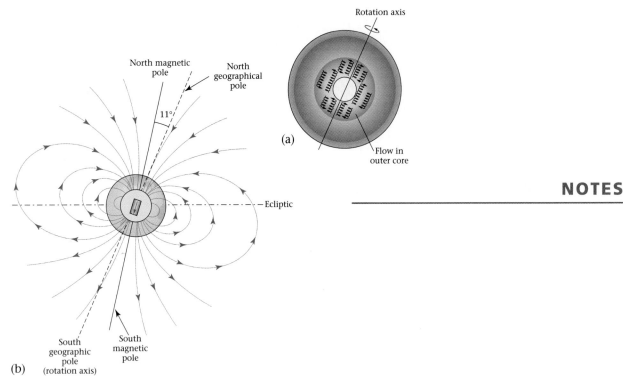

FIGURE 3.8 (a) The convective flow of liquid iron alloy in the Earth's outer core. (b) Earth's magnetism creates magnetic lines of force in space.

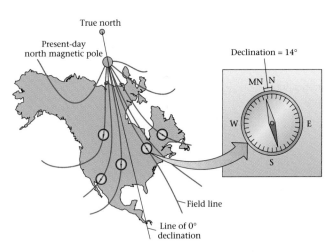

FIGURE 3.9 The projection of magnetic field lines in North America at present.

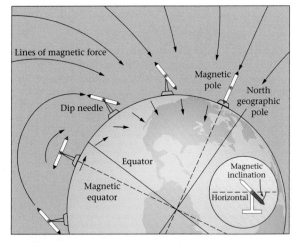

FIGURE 3.10 An illustration of magnetic inclination.

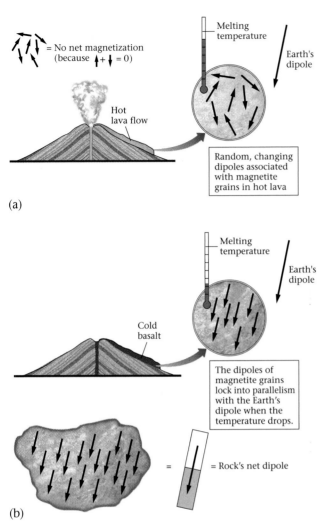

(a)

(b)

FIGURE 3.11 The formation of paleomagnetism. (a) At high temperatures greater than 350°–550°C. (b) As the sample cools to below 350°–550°C.

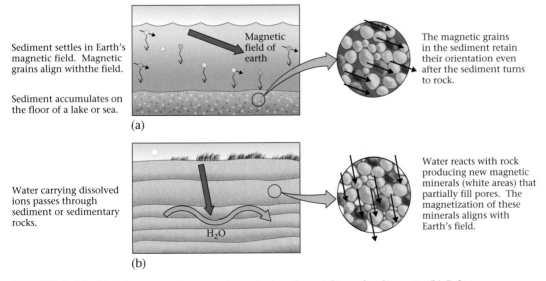

FIGURE 3.12 (a) Paleomagnetism can form during the settling of sediments. (b) Paleomagnetism can also form when iron-bearing minerals precipitate out of groundwater passing through sediment.

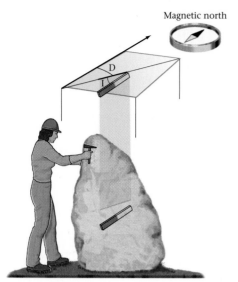

FIGURE 3.13 A rock sample can maintain paleomagnetization for millions of years.

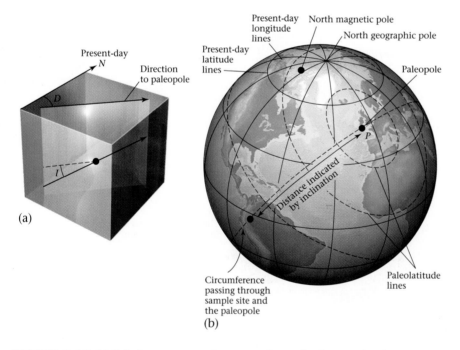

(a)

(b)

FIGURE 3.14 (a) Relative to present-day magnetic north, the sample of Figure 3.13 has a declination angle D in map view, and an inclination angle I in a vertical plane. (b) The paleomagnetic dipole preserved in the rock indicates that, relative to the sample site, the north magnetic pole sat at point P when the rock formed.

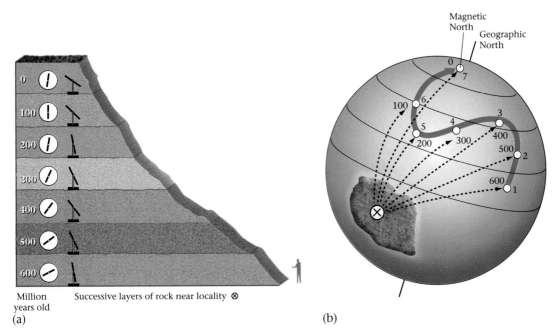

Million
years old   Successive layers of rock near locality ⊗
(a)   (b)

FIGURE 3.15 (a) A cliff at location X exposes a succession of dated lava flows. (b) The succession of paleopole positions through time for location X defines the polar-wander path for the location.

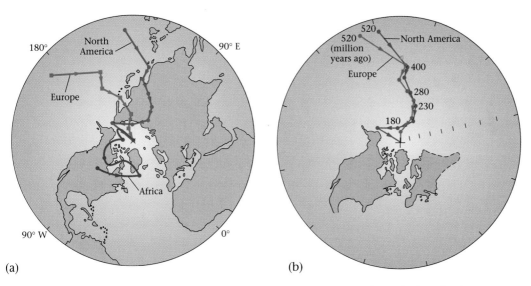

(a)   (b)

FIGURE 3.16 (a) Apparent polar-wander paths of North America, Europe, and Africa for the past several hundred million years. (b) The apparent polar-wander paths for North America and Europe would have coincided with each other from about 280 to 180 million years ago.

NOTES

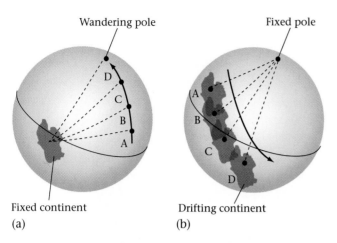

(a)

(b)

FIGURE 3.17 The two alternative explanations for an apparent polar-wander path. (a) A "true polar-wander" model. (b) A continental drift model.

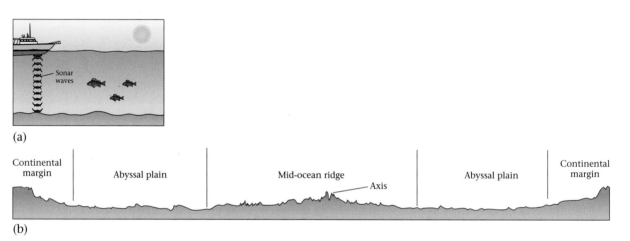

(a)

(b)

FIGURE 3.18 (a) To make a bathymetric profile, researchers use sonar. (b) An east-west bathymetric profile of the Atlantic Ocean.

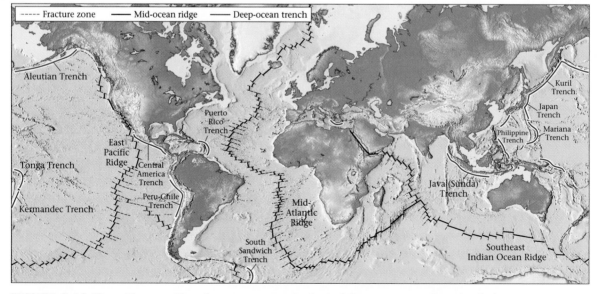

FIGURE 3.19 The mid-ocean ridges, fracture zones, and principal deep-ocean trenches of today's oceans.

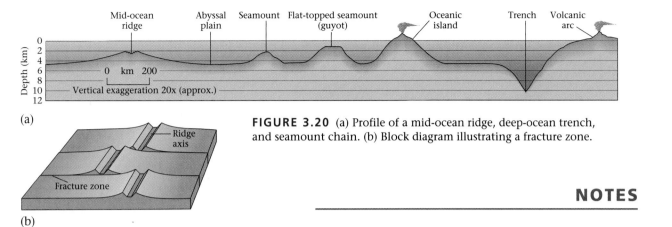

(a)

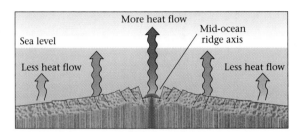

(b)

FIGURE 3.20 (a) Profile of a mid-ocean ridge, deep-ocean trench, and seamount chain. (b) Block diagram illustrating a fracture zone.

NOTES

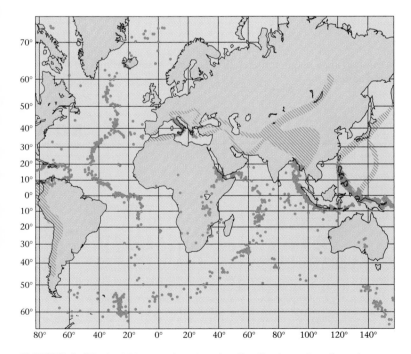

FIGURE 3.21 In a mid-ocean ridge, heat from the mantle flows up through the crust; heat flow decreases away from the ridge axis.

FIGURE 3.22 A 1953 map showing the distribution of earthquake locations in the ocean basins.

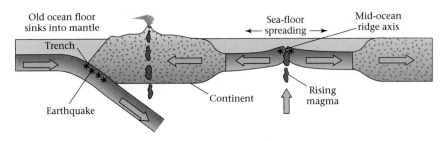

FIGURE 3.23 Harry Hess's basic concept of sea-floor spreading.

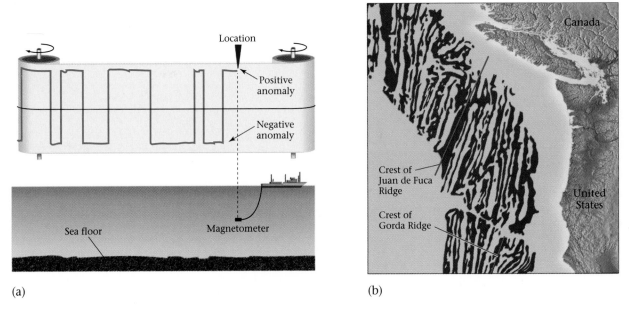

(a)

(b)

FIGURE 3.24 (a) A ship sailing through the ocean dragging a magnetometer. (b) Magnetic anomalies on the sea floor off the northwestern coast of the United States.

NOTES

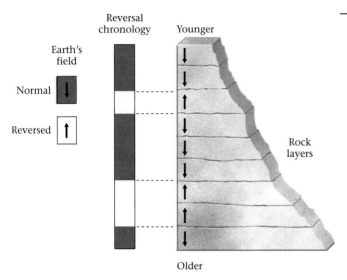

FIGURE 3.25 In a succession of rock layers on land, different flows exhibit different polarity.

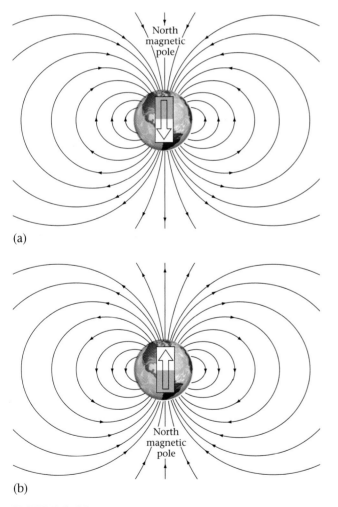

(a)

(b)

FIGURE 3.26 The magnetic field of the Earth has had reversed polarity at various times during Earth history. (a) If the dipole points from north to south, Earth has normal polarity. (b) If the dipole points from south to north, Earth has reversed polarity.

NOTES

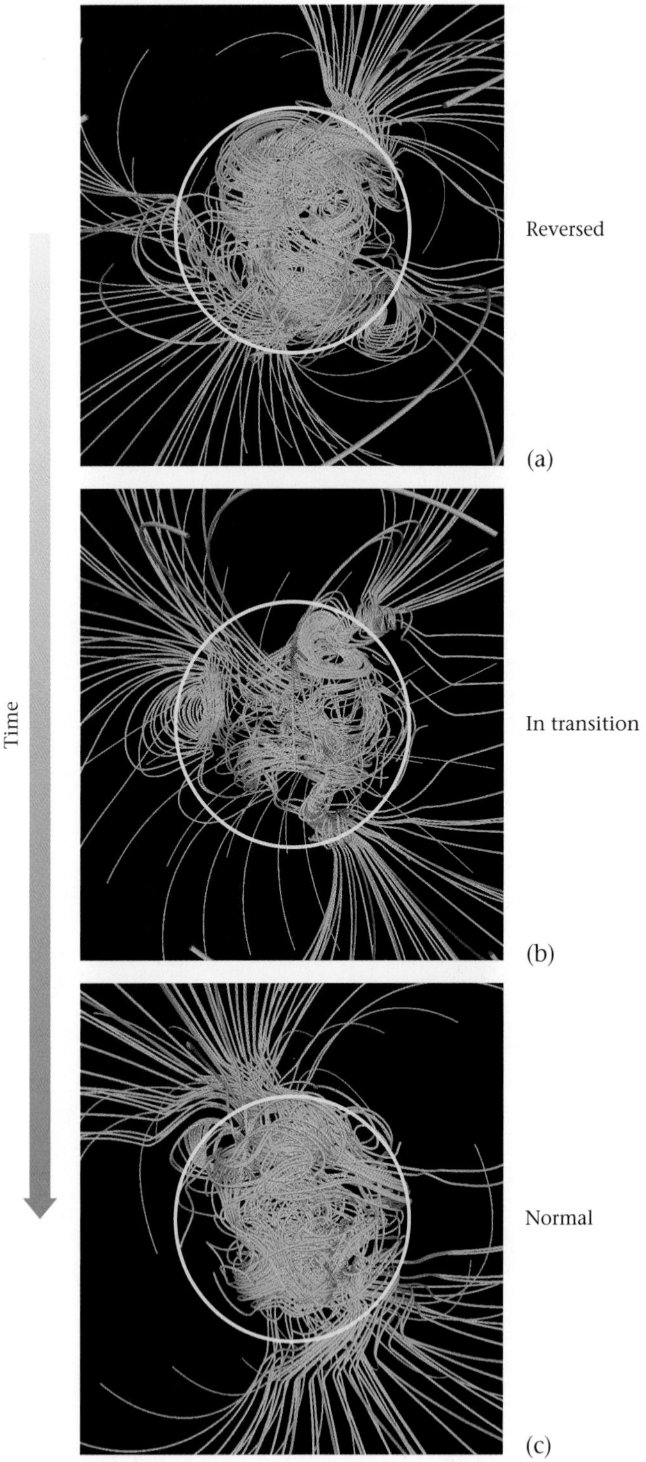

FIGURE 3.27 Earth's magnetic changes during a reversal. (a) Reversed polarity. (b) Polarity during transition. (c) Normal polarity.

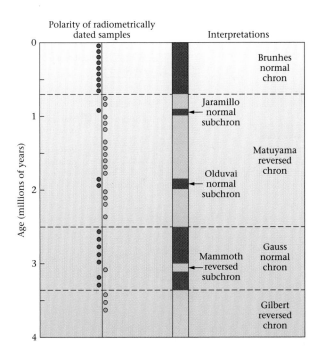

FIGURE 3.28 Radiometric dating of lava flows.

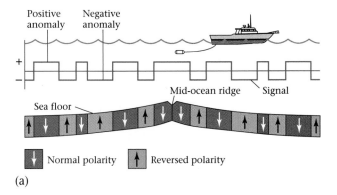

(a)

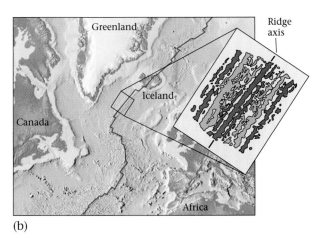

(b)

FIGURE 3.29 (a) The explanation of marine anomalies. (b) The symmetry of the magnetic anomalies measured across the Mid-Atlantic Ridge south of Iceland.

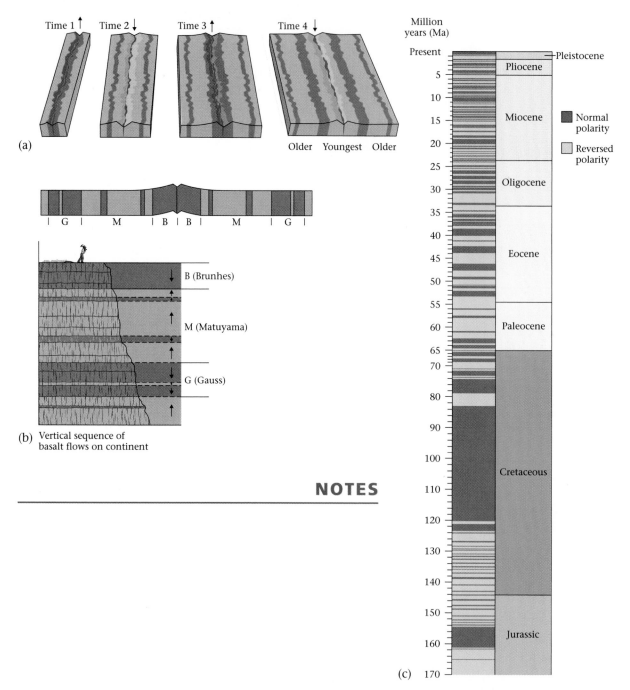

FIGURE 3.30 (a) The progressive development of stripes of alternating polarity in the ocean floor. (b) The observed stripes correlate with the polarity chrons and subchrons measured in lava flows on land. (c) The reversal chronology for the last 170 million years, based on marine magnetic anomalies.

CHAPTER 4 The Way the Earth Works: Plate Tectonics

Time 1: A "load" is placed on top of the lithosphere.
Time 2: The weight of the load pushes down. The lithosphere bends and its base moves down. The plastic asthenosphere flows out of the way.
Load
Lithosphere
Asthenosphere
Load
Bend
Bend
Flow
Flow
Flow
(a)
(not to scale)
Cork
Oak
Pine
Oak
Water (fluid)
Pressure is constant along this line.
(b)
Continental lithosphere (thicker)
Oceanic lithosphere (thinner)
Crust
Moho
Lithosphere
Lithospheric mantle (rigid)
Asthenosphere (plastic)
(c)

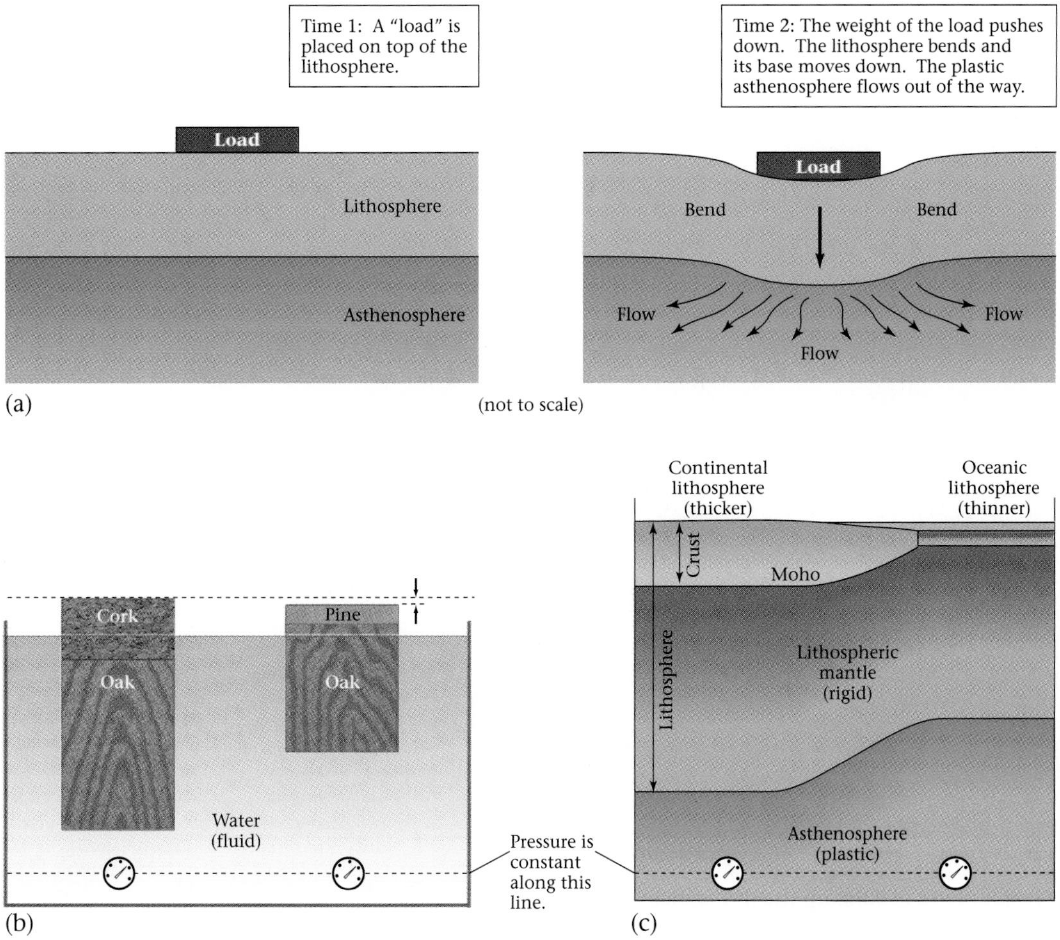

FIGURE 4.1 (a) Lithosphere bends when a load is placed on it, whereas asthenosphere flows. (b) Continental lithosphere as a thick oak block (lithospheric mantle) overlaid by a layer of cork (continental crust), and oceanic lithosphere as a thinner block of oak overlaid by a layer of pine (oceanic crust). (c) Lithosphere, like the wood blocks, floats on the asthenosphere.

NOTES

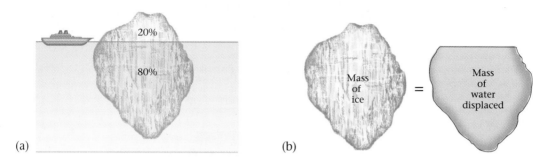

FIGURE 4.2 Archimedes' principle of buoyancy. (a) Ice is less dense than water, so it is buoyant. (b) The ice sinks until the total mass of the water displaced equals the total mass of the whole iceberg.

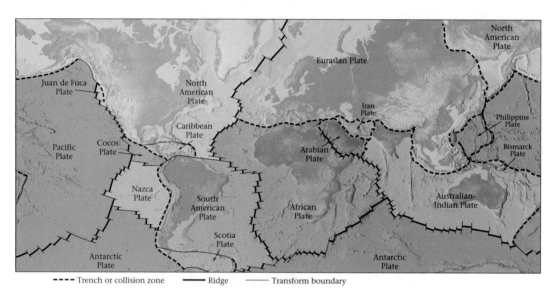

---- Trench or collision zone ——— Ridge ——— Transform boundary

FIGURE 4.3 The major plates making up the lithosphere.

NOTES

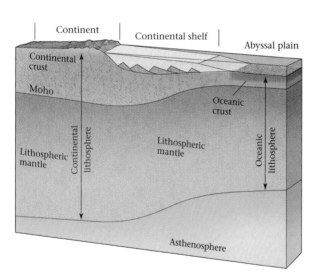

FIGURE 4.4 Block diagram of a passive margin.

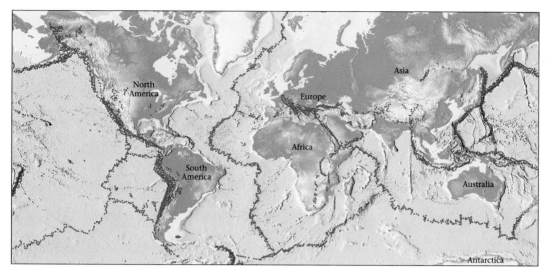

FIGURE 4.5 The locations of most earthquakes fall in distinct bands, or belts.

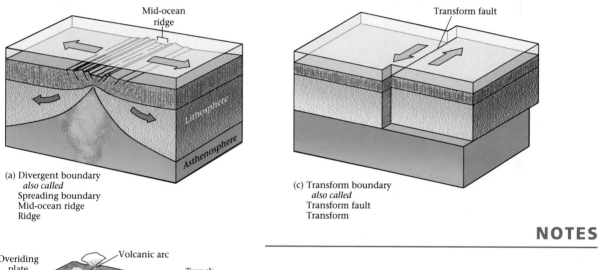

(a) Divergent boundary
also called
Spreading boundary
Mid-ocean ridge
Ridge

(c) Transform boundary
also called
Transform fault
Transform

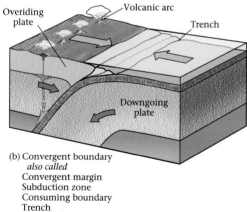

(b) Convergent boundary
also called
Convergent margin
Subduction zone
Consuming boundary
Trench

NOTES

FIGURE 4.6 Geologists recognize three types of plate boundaries. (a) A divergent boundary. (b) A convergent boundary. (c) A transform boundary.

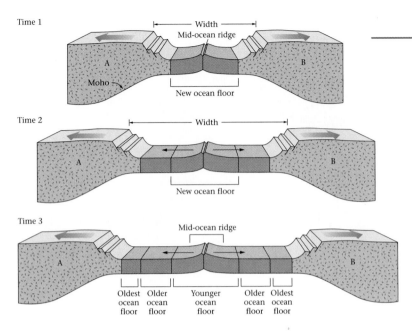

Time 1

Time 2

Time 3

FIGURE 4.7 Sketches depicting successive stages in sea-floor spreading along a divergent boundary.

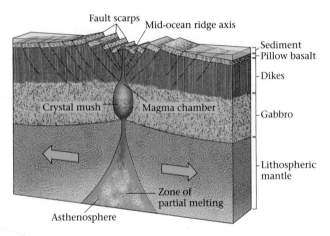

FIGURE 4.8 How new lithosphere forms at a mid-ocean ridge.

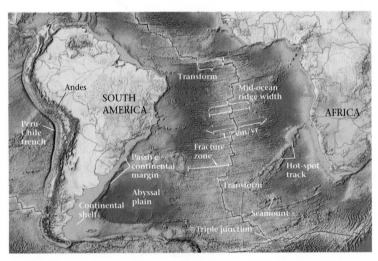

FIGURE 4.9 A map showing the bathymetry of the mid-Atlantic Ridge in the South Atlantic Ocean.

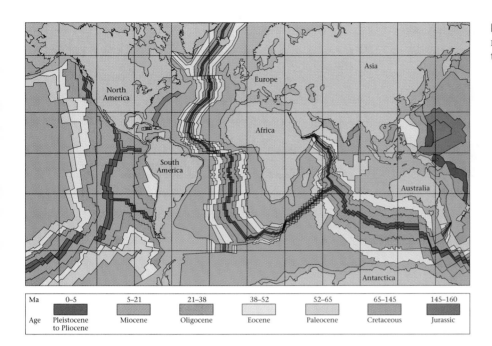

FIGURE 4.11 This map of the world shows the age of the sea floor.

Ma	0–5	5–21	21–38	38–52	52–65	65–145	145–160
Age	Pleistocene to Pliocene	Miocene	Oligocene	Eocene	Paleocene	Cretaceous	Jurassic

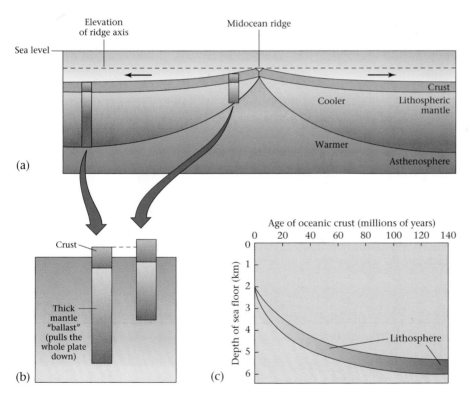

FIGURE 4.12 (a) As sea floor ages, the dense lithospheric mantle thickens. (b) Like the ballast of a ship, thicker lithosphere sinks deeper into the mantle. (c) The thickness of the lithosphere and the depth of the sea floor both increase as a plate moves away from the ridge and grows older.

NOTES

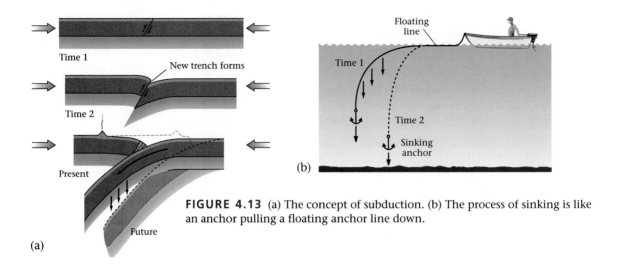

FIGURE 4.13 (a) The concept of subduction. (b) The process of sinking is like an anchor pulling a floating anchor line down.

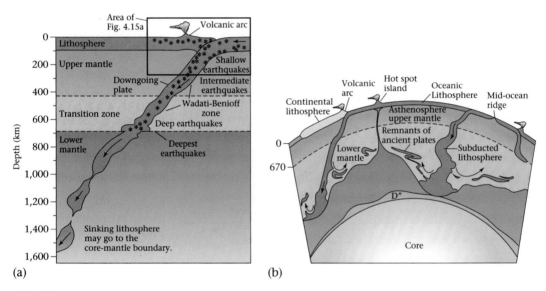

FIGURE 4.14 (a) The Wadati-Benioff zone is a band of earthquakes that occur in subducted oceanic lithosphere. (b) A model illustrating the ultimate fate of subducted lithosphere.

NOTES

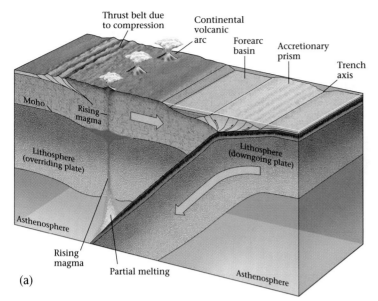

(a)

(b)

FIGURE 4.15 (a) This model shows the geometry of subduction along an active continental margin. (b) The action of a bulldozer pushing snow or soil is similar to the development of an accretionary prism.

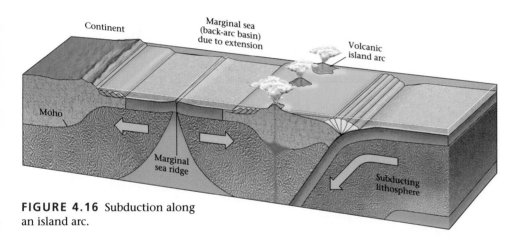

FIGURE 4.16 Subduction along an island arc.

NOTES

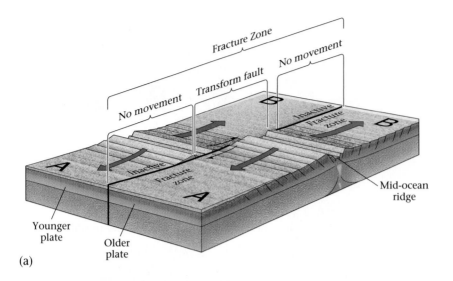

(a)

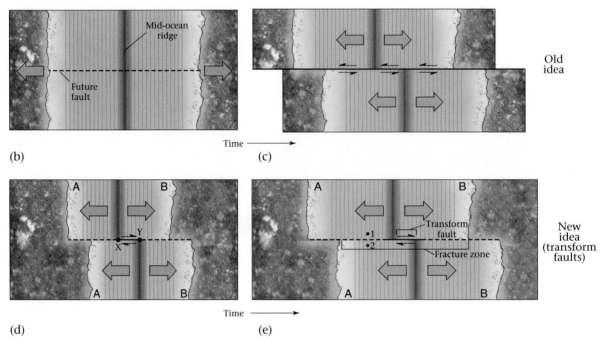

FIGURE 4.17 (a) Fracture zone. (b) An incorrect interpretation of an oceanic fracture zone. (c) After slip on the fault. (d) Wilson's correct interpretation. (e) Even though the ocean grows, the transform fault can stay the same length.

NOTES

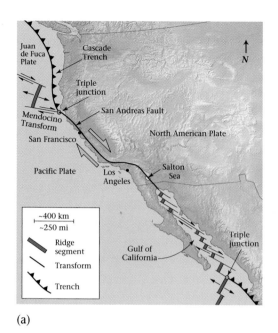

(a)

(b)

FIGURE 4.19 (a) The San Andreas Fault is a transform plate boundary between the Pacific Plate to the west and the North American Plate to the east. (b) The San Andreas Fault where it cuts across a dry landscape.

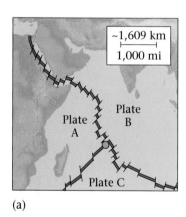

(a)

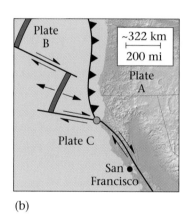

(b)

FIGURE 4.20 (a) A ridge-ridge-ridge triple junction (at the dot). (b) A trench-transform-transform triple junction (at the dot).

NOTES

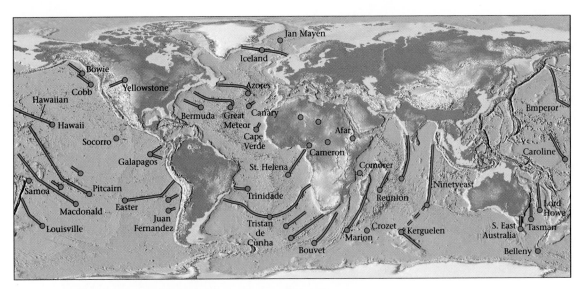

FIGURE 4.21 The dots represent the locations of selected hot-spot volcanoes.

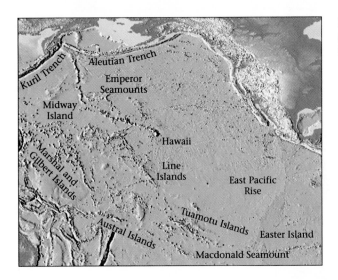

FIGURE 4.22 Bathymetric map showing hot-spot tracks in the Pacific Ocean.

NOTES

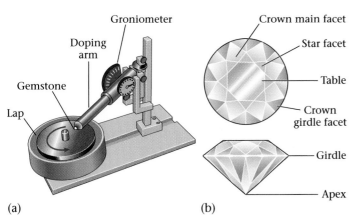

(a)   (b)

FIGURE 5.28 The shiny faces on gems in jewelry are made by a faceting machine. (a) In this faceting machine, the gem is held against the face of the spinning lap. (b) Top and side views show the many facets of a brilliant-cut diamond, and names for different parts of the stone.

INTERLUDE A   *Rock Groups*

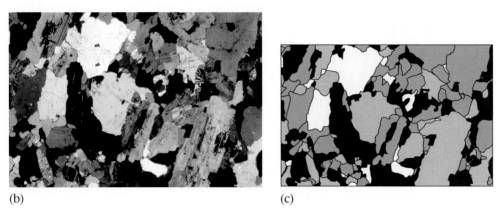

(b)   (c)

FIGURE A.2 (b) A photomicrograph (a photo taken through a microscope) shows the texture of granite is different from that of sandstone. In granite, the grains interlock with one another, like pieces of a jigsaw puzzle. (c) An artist's sketch emphasizes the irregular shapes of grains and how they interlock.

NOTES

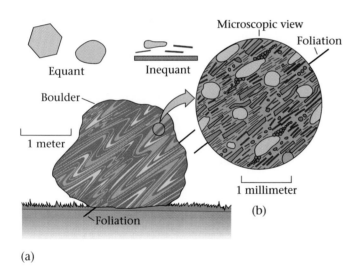

(a)

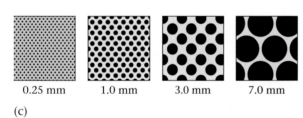

(c)

FIGURE A.5 (a) This boulder of metamorphic rock is an aggregate of mineral grains. (b) At high magnification, we can see that the rock consists of both equant and inequant grains. The inequant grains are aligned parallel to one another. (c) Using this comparison chart, we can measure the size of the grains, in millimeters.

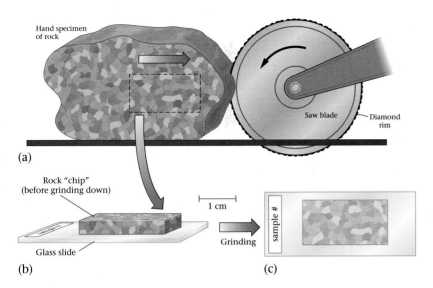

FIGURE A.7 Preparing a thin section.

NOTES

CHAPTER 6 *Up from the Inferno: Magma and Igneous Rocks*

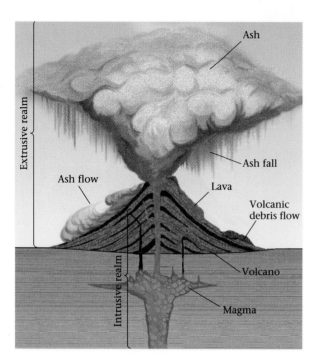

FIGURE 6.2 Extrusive igneous rocks form above the Earth's surface; intrusive rocks develop below.

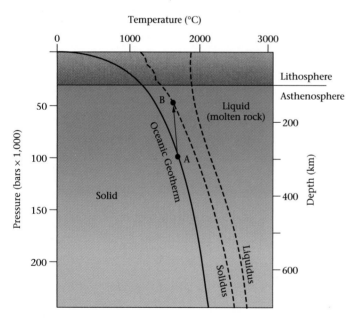

FIGURE 6.3 The graph plots the Earth's geotherm (solid line), as well as the "liquidus" and "solidus" (dashed lines) for peridotite, the ultramafic rock that makes up the mantle.

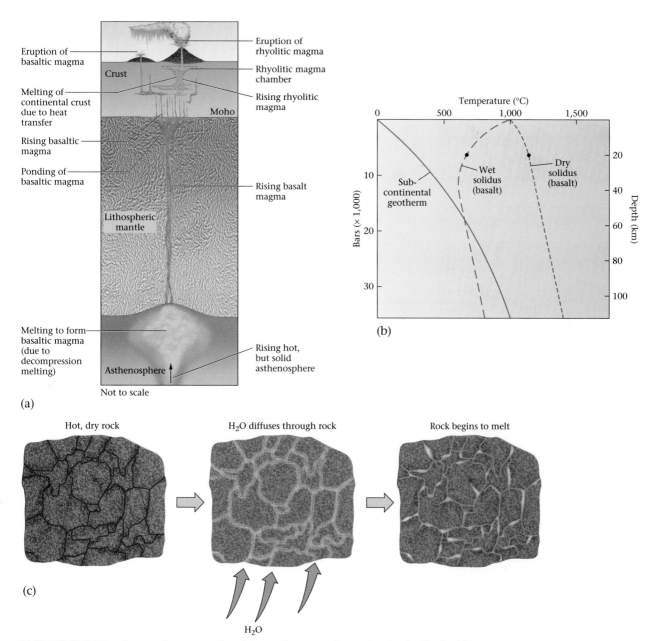

FIGURE 6.4 The three main causes of melting and magma formation in the Earth. (a) Decompression melting. (b) The addition of volatiles decreases the melting temperature. (c) Melting as a result of the addition of volatiles.

NOTES

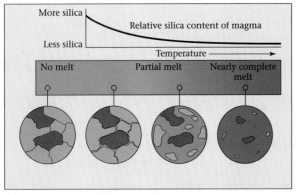

(a)

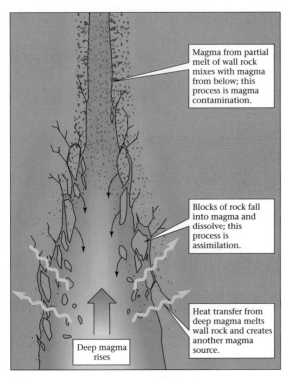

(b)

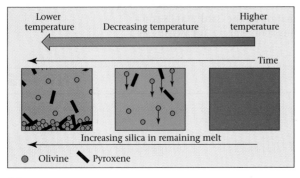

(c)

FIGURE 6.5 (a) The concept of partial melting. (b) The concept of magma contamination.

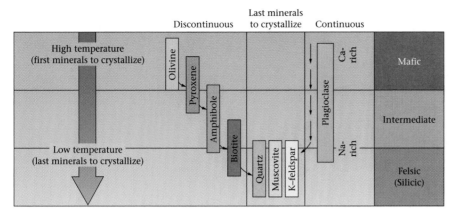

FIGURE 6.6 Bowen's reaction series.

(a)

(b)   Not to scale

FIGURE 6.7 The behavior of erupting lava reflects its viscosity. (a) Viscous lava forms a blob or mound-like dome. (b) Nonviscous lava spreads out in a thin flow.

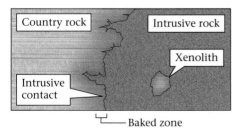

FIGURE 6.9 An intrusive contact.

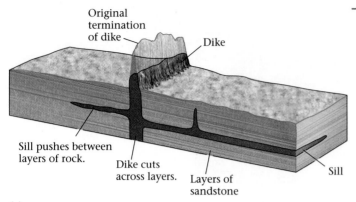

Original termination of dike

Dike

Sill pushes between layers of rock.

Dike cuts across layers.

Layers of sandstone

Sill

(a)

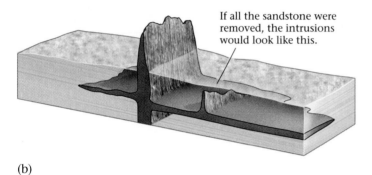

If all the sandstone were removed, the intrusions would look like this.

(b)

FIGURE 6.10 (a) Dikes and sills are vertical or horizontal bands, respectively, on the face of an outcrop. (b) If we were to strip away the surrounding rock, dikes would look like walls, and sills would look like tabletops.

FIGURE 6.11 (a) A basalt dike looks like a black stripe painted on an outcrop of granite (here, in Arizona). (b) At this ancient volcano called Shiprock, in New Mexico, ash and lava flows have eroded away, leaving a "volcanic neck."
(c) Shiprock was once in the interior of a volcano or below a volcano. (d) These Precambrian dikes exposed in the Canadian Shield formed when the region underwent stretching over a billion years ago. (e) This dark sill, exposed on a cliff in Antarctica, is basalt; the white rock is sandstone. (f) This geologist's sketch shows the cliff face as viewed face on. (g) Map showing Cenozoic dikes in the United Kingdom and Ireland.

(a)

(b)

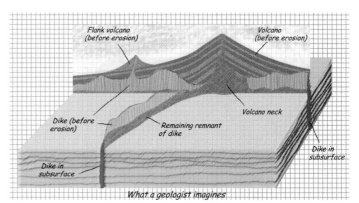
(c)

(d)

(e)

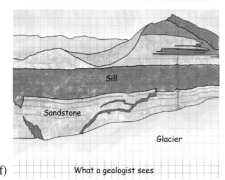
(f)

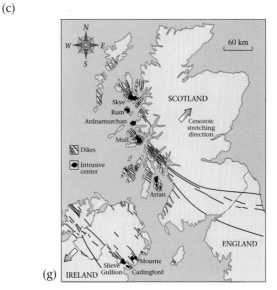
(g)

NOTES

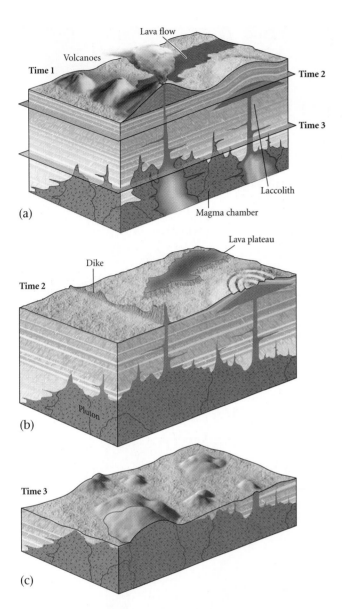

FIGURE 6.12 (a) While a volcano is active, a magma chamber exists underground; dikes, sills, and laccoliths intrude; and lava and ash erupt at the surface. (b) Later, the bulbous magma chamber freezes into a pluton. The soft parts of the volcano erode. (c) With still more erosion, volcanic rocks and shallow intrusions are removed, and we see plutonic intrusive rocks.

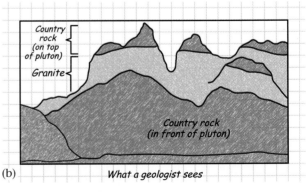

FIGURE 6.13 (a) Torres del Paines, a spectacular group of mountains in southern Chile. (b) A geologist's sketch, labeling the two major rock units.

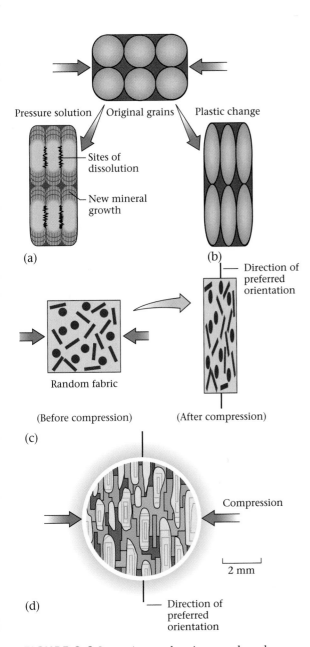

(a)

(b)

Direction of preferred orientation

Random fabric

(Before compression)

(After compression)

(c)

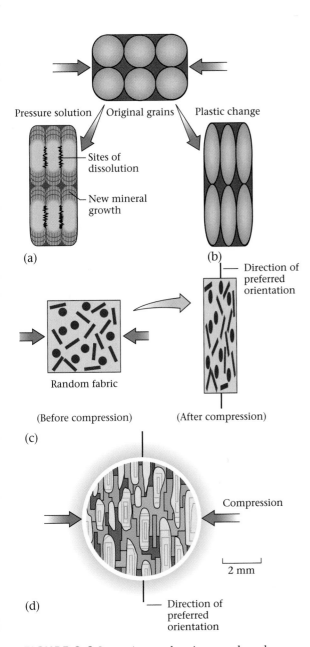

Compression

2 mm

(d)

Direction of preferred orientation

FIGURE 8.6 Squeezing or shearing a rock under metamorphic conditions can result in preferred mineral orientation, in four ways. (a) Plastic deformation. (b) Grains can undergo pressure solution. (c) Inequant grains distributed through a soft matrix. (d) Inequant grains can grow with a preferred orientation.

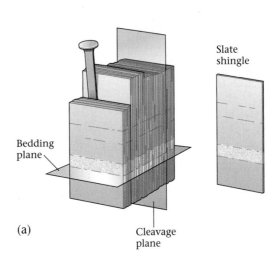

Slate shingle

Bedding plane

(a)

Cleavage plane

FIGURE 8.8 (a) A block of rock with slaty cleavage splits along cleavage planes into thin sheets.

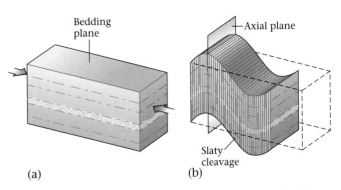

Bedding plane

Axial plane

(a)

(b)

Slaty cleavage

FIGURE 8.9 (a) The end-on compression of a bed will create slaty cleavage at an angle perpendicular to the bedding. (b) Commonly, the rock folds (bends into curves) at the same time cleavage forms.

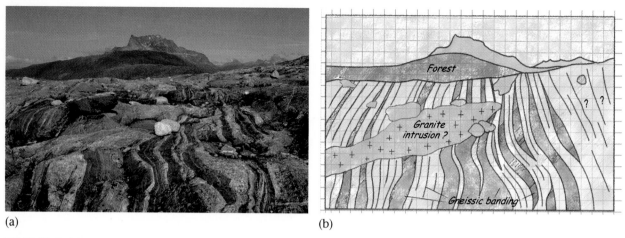

(a) (b)

FIGURE 8.11 (a) A glaciated surface exposing gneiss contains alternating bands of light-colored and dark-colored minerals. (b) A geologist's interpretation of the outcrop emphasizes the banding.

NOTES

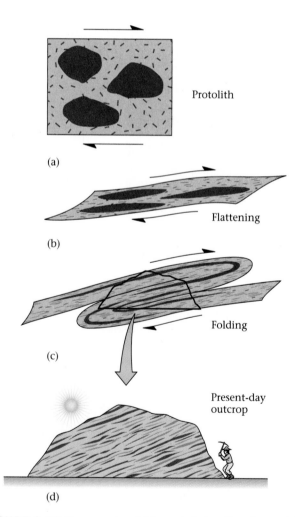

(a) Protolith

(b) Flattening

(c) Folding

(d) Present-day outcrop

FIGURE 8.12 The ways in which gneissic banding forms. (a) The protolith contains patches that are more mafic than the surrounding felsic rock. (b) Shear stretches and flattens the rock. (c) The layer is folded back on itself in response to continued shear. (d) A present-day outcrop of this rock displays mafic bands separated by felsic bands.

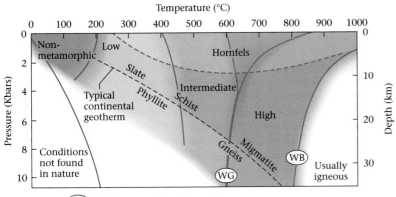

Melting curve for wet granite (WG)

Melting curve for wet basalt (WB)

PROTOLITH	LOW	INTERMEDIATE	HIGH	PARTIAL MELTING
Basalt (mafic)	Greenschist	Amphibolite	Mafic granulite	

	Zeolite				
	Chlorite				
		Epidote			
	No Al	Amphibole	Al		
		Garnet			
		Pyroxene			

PROTOLITH	LOW	INTERMEDIATE	HIGH	PARTIAL MELTING
Shale (pelitic)	Slate Phyllite — Schist — Gneiss —			Migmatite

Clay				
	Chlorite			
		Quartz/Feldspar		
		Muscovite		
		Biorite		
		Garnet		
		Staurolite		
		Kyanite		
			Sillimanite	

FIGURE 8.17 The concept of metamorphic grade. (a) A graph showing the approximate conditions of various grades. (b) The minerals that form in a given rock depend on grade *and* composition.

NOTES

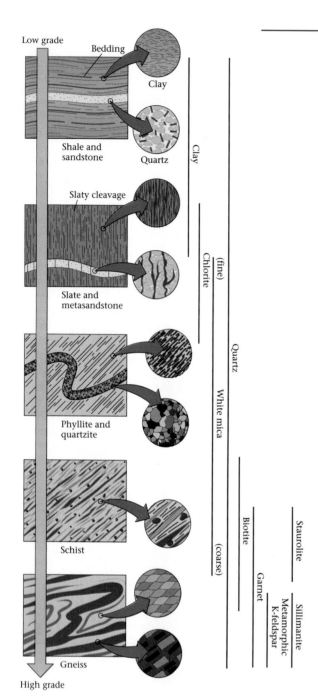

FIGURE 8.18 When shale progressively metamorphoses from low grade to high grade, it first becomes slate, then phyllite, then schist, then gneiss.

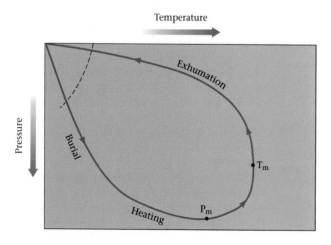

FIGURE 8.19 The metamorphic history of a rock can be portrayed on a graph showing variation in temperature and pressure.

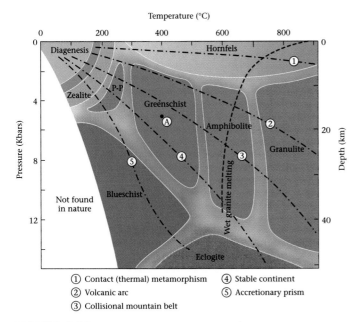

① Contact (thermal) metamorphism   ④ Stable continent
② Volcanic arc   ⑤ Accretionary prism
③ Collisional mountain belt

FIGURE 8.20 The common metamorphic facies.

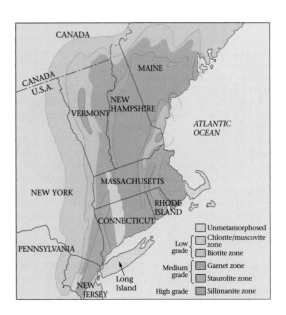

FIGURE 8.21 Metamorphic zones as portrayed on a map of New England (U.S.A.).

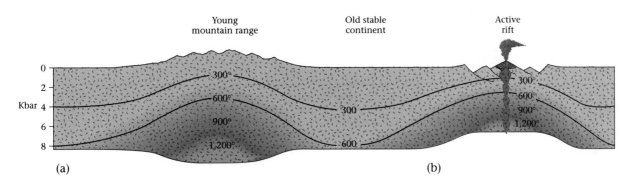

FIGURE 8.22 The change in temperature with depth at different locations in a continent.

NOTES

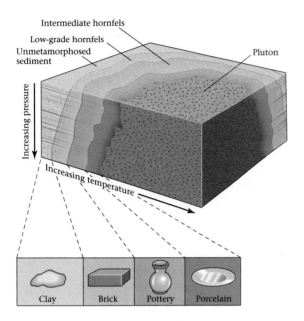

FIGURE 8.23 A metamorphic aureole bordering an igneous intrusion.

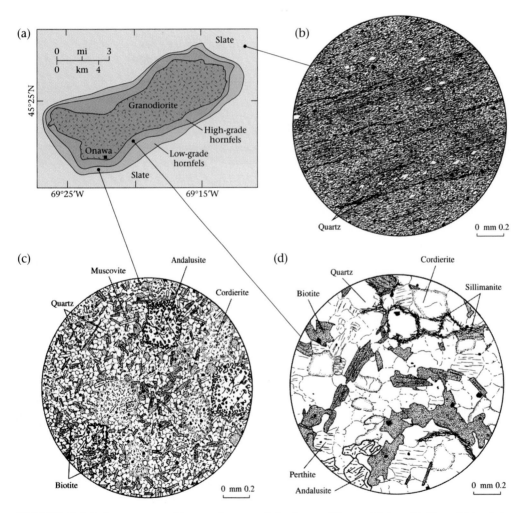

FIGURE 8.24 The metamorphic aureole around the Onawa Pluton, Maine. (a) The width of the preserved aureole varies with location (b) Far from the pluton, the country rock is a slate composed of aligned clay and very fine quartz. (c) In the low-grade part of the aureole, a totally new hornfels texture has formed. (d) In the high-grade part of the aureole, the hornfels is much coarser and contains different minerals.

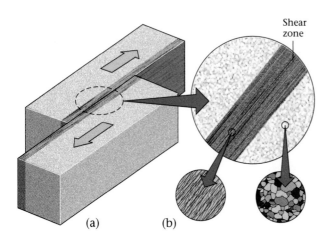

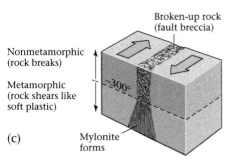

FIGURE 8.25 Dynamic metamorphism along a fault zone. (a) Note the band of sheared rock on either side of the slip surface. (b) The rock outside the shear zone has a different texture from the rock inside. (c) The block formed in *(a)* must have developed at a depth where metamorphic conditions exist, so that mylonite forms.

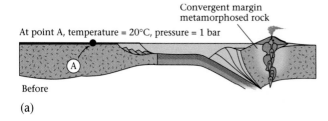

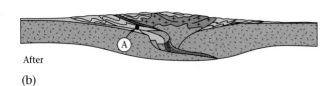

FIGURE 8.26 (a) Metamorphism occurs where there is plutonic activity along a convergent boundary. (b) The sedimentary rock that lay at the top of a passive margin gets carried to great depth in a continental collision that leads to mountain building.

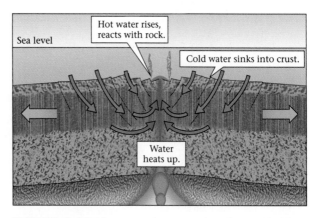

FIGURE 8.27 Along a mid-ocean ridge, the circulation of hydrothermal fluids in response to igneous activity along mid-ocean ridges causes metamorphism of basalt in the oceanic crust.

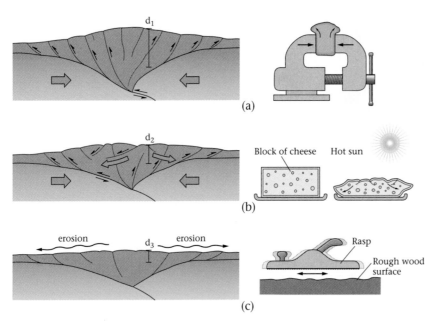

FIGURE 8.28 Three geologic phenomena, together, can contribute to exhumation in a collisional mountain belt. (a) Collision squeezes rock in the mountain belt upward, like dough pressed in a vise. (b) Eventually, the crust beneath the mountain range becomes warm and weak, so the mountain belt collapses, like a block of soft cheese placed in the hot sun. (c) Throughout the history of the mountain belt, erosion grinds rock off the surface and removes it, much like a giant rasp.

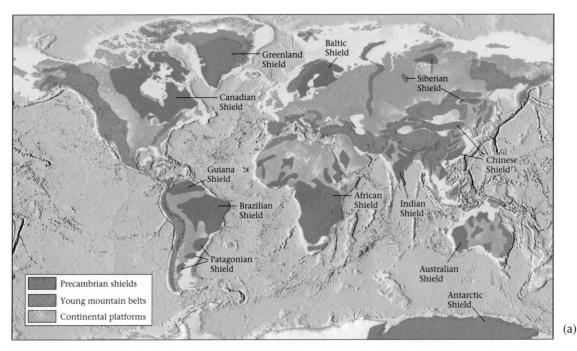

FIGURE 8.29 (a) The distribution of shield areas (exposed Precambrian metamorphic and igneous rock) on the Earth.

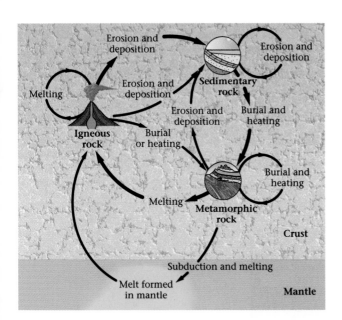

INTERLUDE B *The Rock Cycle*

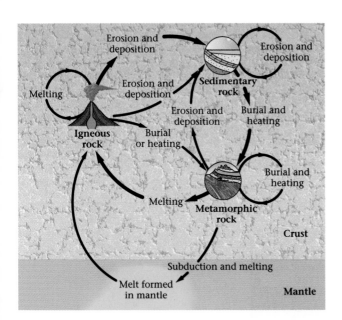

FIGURE B.1 The stages of the rock cycle, showing various alternative pathways.

NOTES

TIME 1
(a)

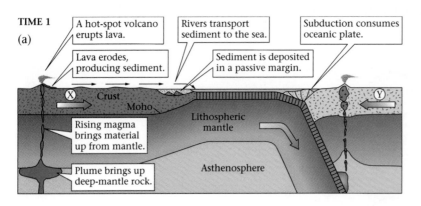

A hot-spot volcano erupts lava.

Lava erodes, producing sediment.

Rivers transport sediment to the sea.

Sediment is deposited in a passive margin.

Subduction consumes oceanic plate.

X   Crust
Moho
Lithospheric mantle

Rising magma brings material up from mantle.

Plume brings up deep-mantle rock.

Asthenosphere

Y

TIME 2
(b)

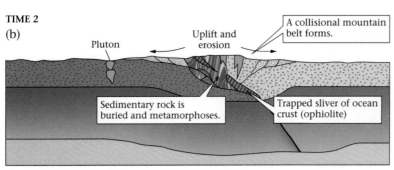

Pluton

Uplift and erosion

A collisional mountain belt forms.

Sedimentary rock is buried and metamorphoses.

Trapped sliver of ocean crust (ophiolite)

▨ Metamorphic rock   ☐ Sediment eroded from mountains

TIME 3
(c)

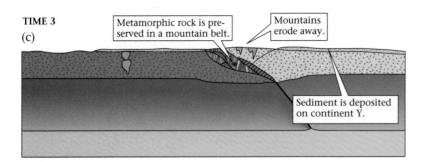

Metamorphic rock is preserved in a mountain belt.

Mountains erode away.

Sediment is deposited on continent Y.

TIME 4
(d)

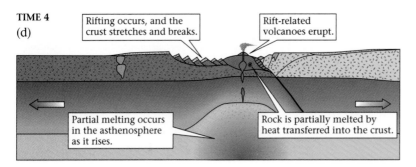

Rifting occurs, and the crust stretches and breaks.

Rift-related volcanoes erupt.

Partial melting occurs in the asthenosphere as it rises.

Rock is partially melted by heat transferred into the crust.

FIGURE B.2 (a) At the beginning of the rock cycle (time 1), atoms, originally making up peridotite in the mantle, rise in a mantle plume. (b) At time 2, continents X and Y collide, and the shale is buried deeply beneath the resulting mountain range (at the dot). (c) At time 3, the mountain range erodes away and the schist rises but does not reach the surface. (d) At time 4, rifting begins to split the continents apart, and igneous activity occurs again.

NOTES

CHAPTER 9 *The Wrath of Vulcan: Volcanic Eruptions*

(a)

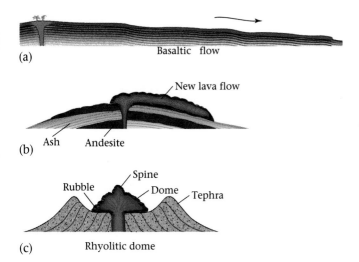

What a geologist imagines

(b)

FIGURE 9.1 (a) Pompeii. (b) What a geologist imagines: When Mt. Vesuvius erupted in 79 C.E.

(a) Basaltic flow

(b) Ash Andesite New lava flow

(c) Rhyolitic dome Rubble Spine Dome Tephra

FIGURE 9.2 The character of a lava flow depends on the viscosity of the lava. (a) A basaltic lava flow is very fluid-like. (b) An andesitic flow is too viscous to travel far. (c) Rhyolitic lava is so viscous that it piles up at the vent in the shape of a dome.

Earth: Portrait of a Planet
W. W. Norton & Company, Inc., 500 Fifth Avenue, New York, N.Y. 10110

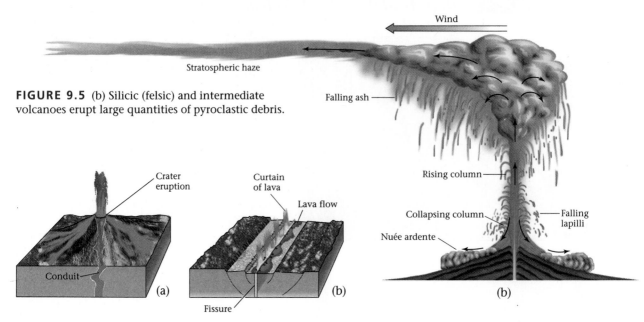

FIGURE 9.5 (b) Silicic (felsic) and intermediate volcanoes erupt large quantities of pyroclastic debris.

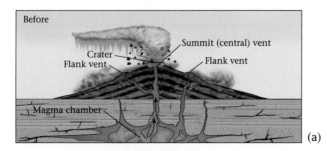

FIGURE 9.9 (a) Some volcanoes erupt out of a circular vent above a tube-shaped conduit. (b) Other volcanoes erupt out of a long crack, called a fissure.

NOTES

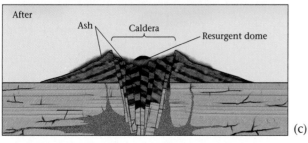

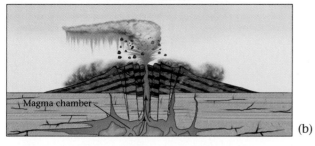

FIGURE 9.10 (a) The plumbing beneath a volcano can be complex. (b) During an eruption, the magma chamber beneath a volcano is inflated with magma. (c) If the eruption drains the magma chamber, the volcano collapses downward to form a circular depression called a caldera.

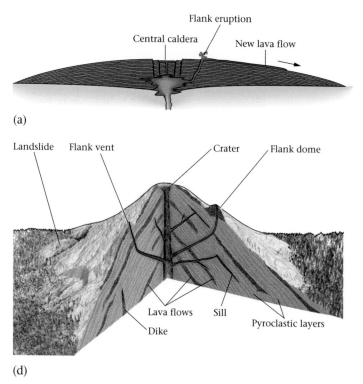

(a)

(d)

FIGURE 9.11 (a) A shield volcano. (d) A composite volcano consists of alternating tephra and lava.

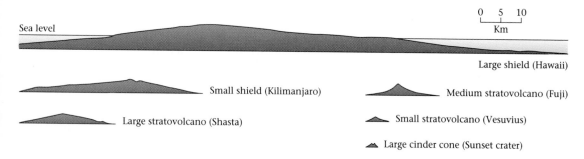

FIGURE 9.12 These profiles emphasize that volcanoes come in different sizes.

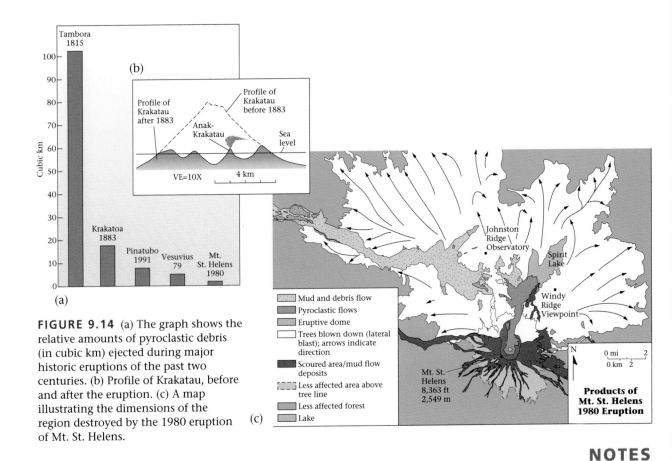

FIGURE 9.14 (a) The graph shows the relative amounts of pyroclastic debris (in cubic km) ejected during major historic eruptions of the past two centuries. (b) Profile of Krakatau, before and after the eruption. (c) A map illustrating the dimensions of the region destroyed by the 1980 eruption of Mt. St. Helens.

NOTES

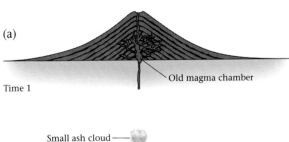

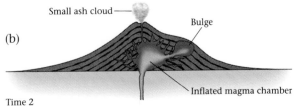

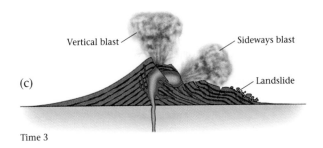

FIGURE 9.15 The eruption of Mt. St. Helens, 1980. (a) Before the eruption, the magma chamber is empty. (b) The magma chamber fills. (c) The weakened north flank suddenly slipped.

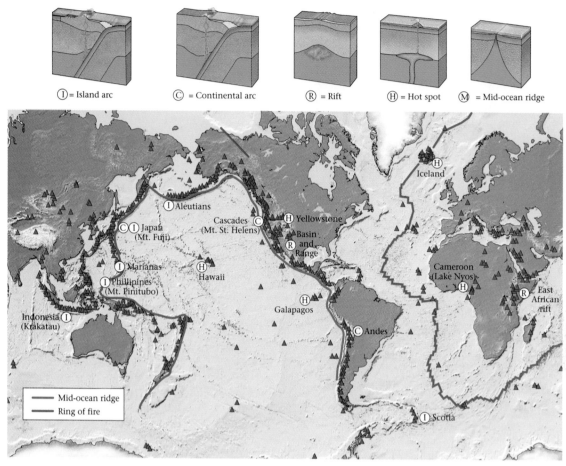

ⓘ = Island arc   ⓒ = Continental arc   Ⓡ = Rift   Ⓗ = Hot spot   Ⓜ = Mid-ocean ridge

FIGURE 9.16 A map showing the distribution of volcanoes around the world, and the basic geologic settings in which volcanoes form, in the context of plate-tectonics theory.

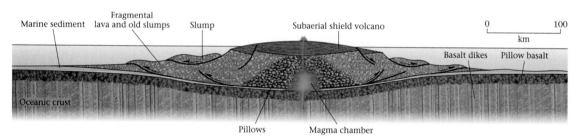

FIGURE 9.17 The inside of an oceanic hot-spot volcano.

NOTES

FIGURE 9.18 (a) Volcanic rocks from hot spots formed in two places in the western United States.

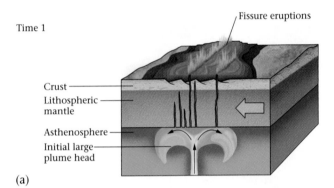

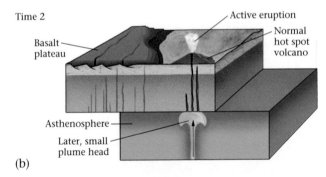

FIGURE 9.19 (a) The lithosphere cracks and rifts above the bulbous head of a plume. (b) Later, the bulbous plume head no longer exists, leaving only a narrower plume stalk.

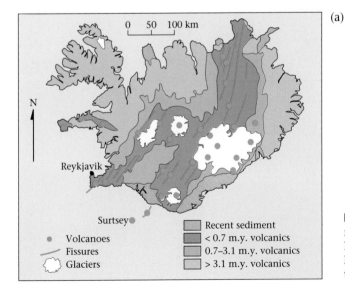

FIGURE 9.21 (a) Iceland consists of volcanic rocks that erupted from a hot spot along the Mid-Atlantic Ridge. (b) The surface of Iceland has dropped down along the faults that bound the central rift.

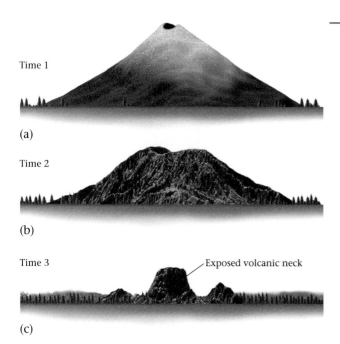

FIGURE 9.22 (a) The shape of an active volcano is defined by the surface of the most recent lava flow or ash fall. (b) A dormant volcano that has been around long enough for the surface to be modified by erosion. (c) An extinct volcano has been so deeply eroded that only the neck of the volcano may remain.

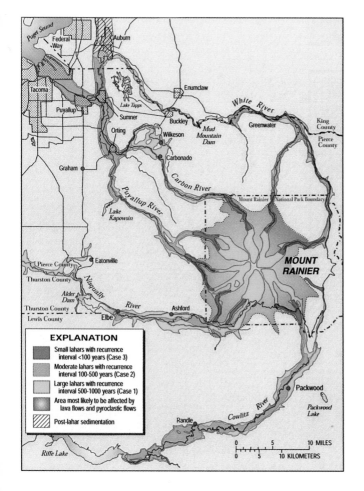

FIGURE 9.25 A danger-assessment map for the Mt. Rainer area.

CHAPTER 10 *A Violent Pulse: Earthquakes*

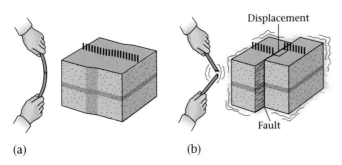

(a)   (b)

FIGURE 10.1 (a) Most earthquakes happen when rock in the ground first bends slightly and then suddenly snaps and breaks, like a stick you flex in your hands. (b) When the crust breaks, sliding suddenly occurs.

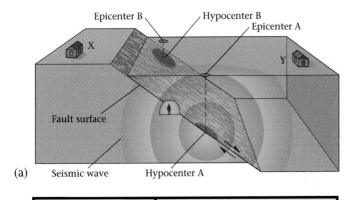

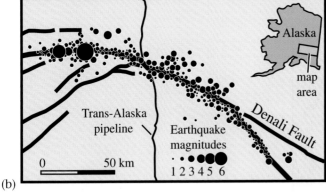

FIGURE 10.4 (a) The energy of an earthquake radiates from the hypocenter (focus). (b) A map of the Denali National Park region, Alaska.

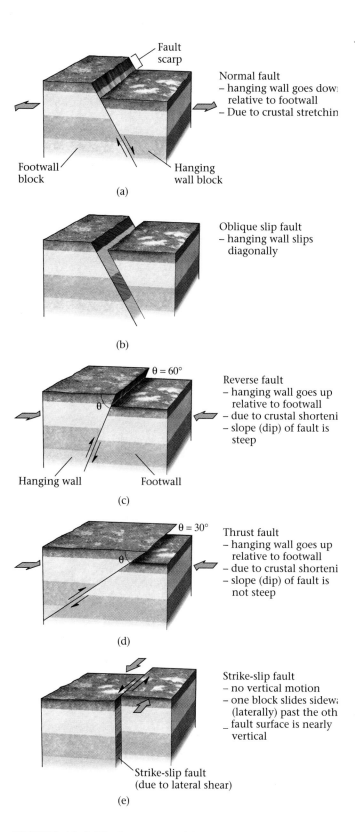

Fault scarp

Normal fault
– hanging wall goes down relative to footwall
– Due to crustal stretching

Footwall block

Hanging wall block

(a)

Oblique slip fault
– hanging wall slips diagonally

(b)

θ = 60°

Reverse fault
– hanging wall goes up relative to footwall
– due to crustal shortening
– slope (dip) of fault is steep

Hanging wall

Footwall

(c)

θ = 30°

Thrust fault
– hanging wall goes up relative to footwall
– due to crustal shortening
– slope (dip) of fault is not steep

(d)

Strike-slip fault
– no vertical motion
– one block slides sideways (laterally) past the other
– fault surface is nearly vertical

Strike-slip fault (due to lateral shear)

(e)

FIGURE 10.5 The basic types of faults. (a) Normal fault, (b) Oblique-slip fault, (c) Reverse fault, (d) Thrust fault, (e) Strike-slip fault.

(a)

(b)

FIGURE 10.6 (a) This wooden fence was built across the San Andreas Fault. (b) The amount the fence was offset indicates the displacement on the fault.

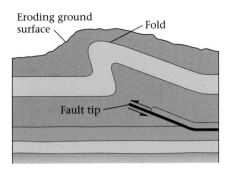

FIGURE 10.7 A hidden (or "blind") fault does not intersect the ground surface.

NOTES

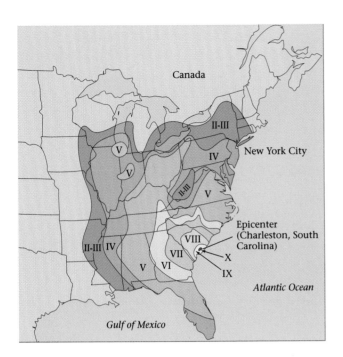

FIGURE 10.17 This map shows the contours of Mercalli intensity for the 1886 Charleston, South Carolina, earthquake.

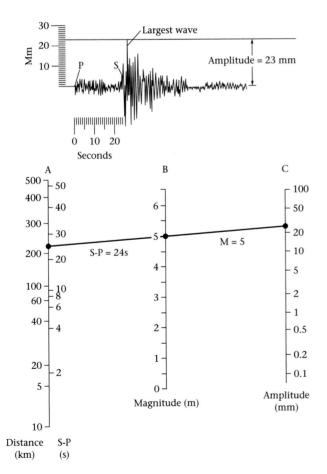

FIGURE 10.18 How to calculate the Richter magnitude from a seismogram.

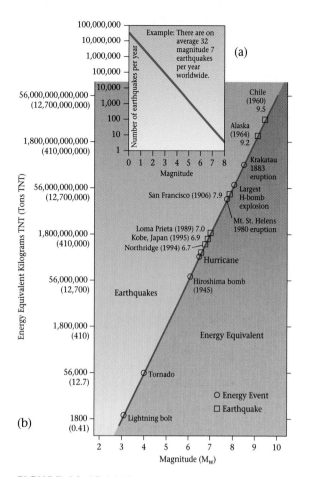

FIGURE 10.19 (a) This graph emphasizes that the energy released by an earthquake increases dramatically with magnitude. (b) This graph illustrates how the number of earthquakes of a given magnitude decreases with increasing magnitude.

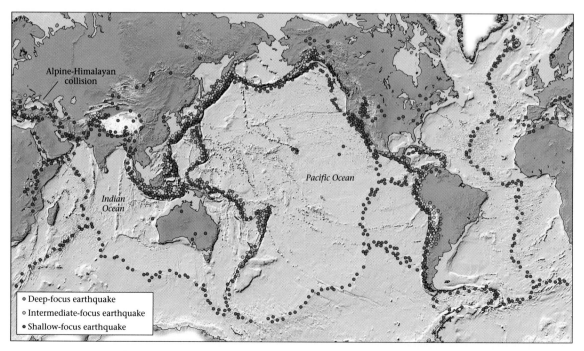

FIGURE 10.20 A simplified map of epicenters shows that most earthquakes occur in distinct belts that define plate boundaries.

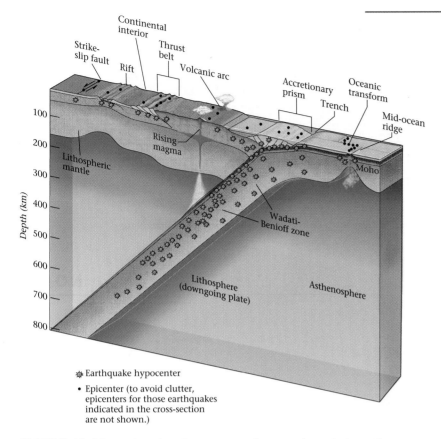

FIGURE 10.21 Earthquakes along oceanic divergent boundaries and transform boundaries are all shallow-focus.

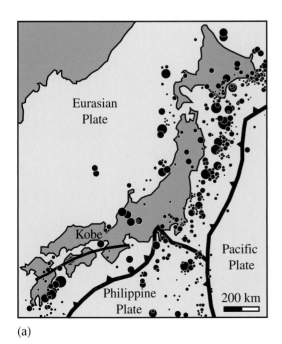

(a)

FIGURE 10.23 (a) Simplified map of earthquake epicenters (black dots) and plate boundaries in Japan. The size of the dots indicates magnitude. Note the location of Kobe.

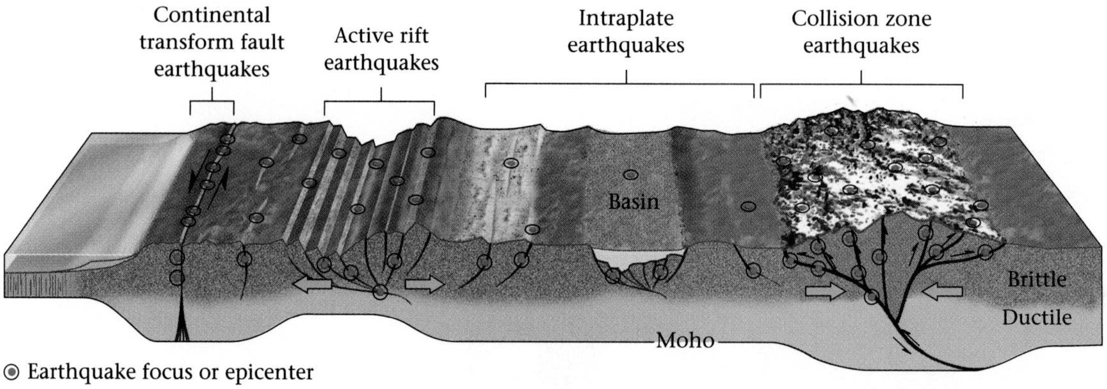

Continental
transform fault
earthquakes

Active rift
earthquakes

Intraplate
earthquakes

Collision zone
earthquakes

Basin

Brittle

Ductile

Moho

⊚ Earthquake focus or epicenter

FIGURE 10.24 Earthquakes occur in continental transform faults, in continental rifts, in intraplate settings, and in collision zones.

NOTES

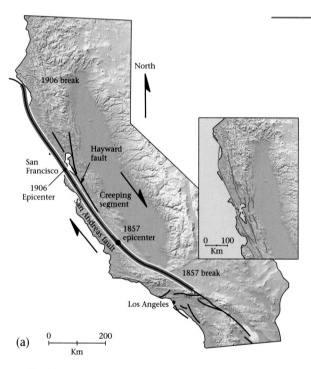

North

1906 break

Hayward
fault

San
Francisco

1906
Epicenter

Creeping
segment

San Andreas fault

1857
epicenter

1857 break

Los Angeles

0   100
Km

(a)   0   200
Km

FIGURE 10.25 (a) Map showing the portion of the San Andreas Fault that ruptured during the 1906 earthquake.

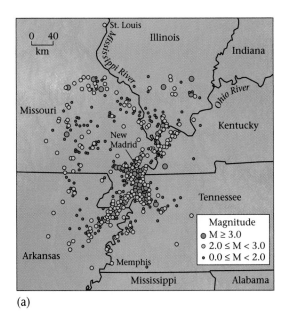

(a)

FIGURE 10.26 (a) The epicenters of earthquakes in the New Madrid, Missouri, area, recorded by modern seismic instruments.

NOTES

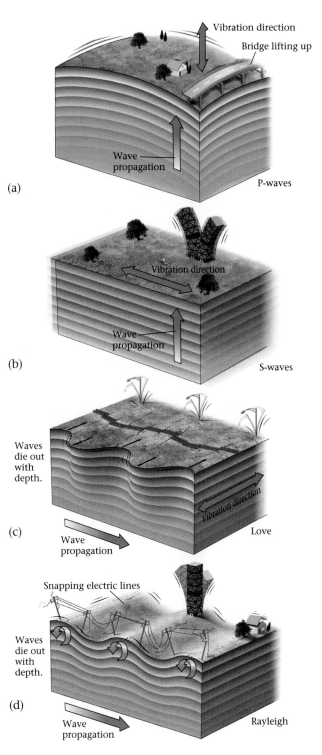

FIGURE 10.27 (a) P-waves (b) S-waves (c) Love waves cause the ground to undulate laterally. (d) Rayleigh waves make the ground undulate in a rolling motion, like the sea surface.

FIGURE 10.28 Shaking causes (a) a concrete-slab building (or bridge) to disconnect and collapse, (b) a building's facade to fall off, (c) a poorly supported bridge to topple, (d) a bridge span to disconnect and collapse, (e) neighboring buildings to collide and shatter (floors inside a tall building may collapse), (f) a concrete-block, brick, or adobe building to crack apart and collapse, (g) a steep cliff to collapse, carrying buildings with it.

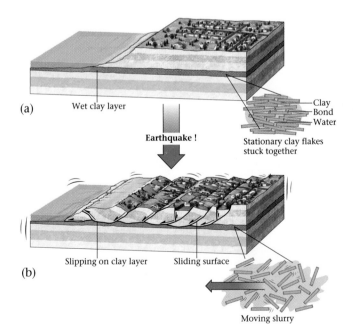

FIGURE 10.31 The Turnagain Heights (Alaska) disaster. (a) Before the earthquake, wet clay packed together in a subsurface layer of compacted but wet mud. (b) Ground shaking caused liquefaction of the wet clay layer in the sediment beneath a housing development.

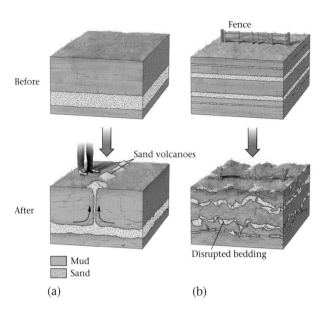

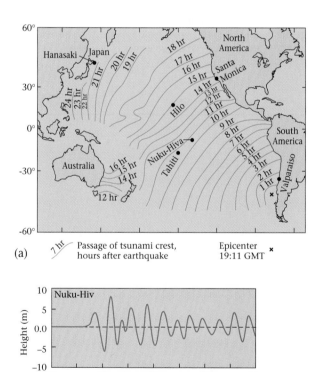

FIGURE 10.32 (a) Ground shaking may cause wet sand buried below the surface to squirt up between cracks in overlying compacted mud. (b) Thin layers of sand interlayered with mud may get disrupted, folded, and broken up as a result of ground shaking.

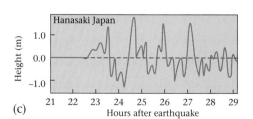

FIGURE 10.34 (a) Map illustrating the successive positions of the tsunami generated by the 1964 Chile earthquake. (b) Tidal gauge at the island of Nuku-Hiv. (c) Tidal gauge at Hanasaki, Japan, twelve hours later.

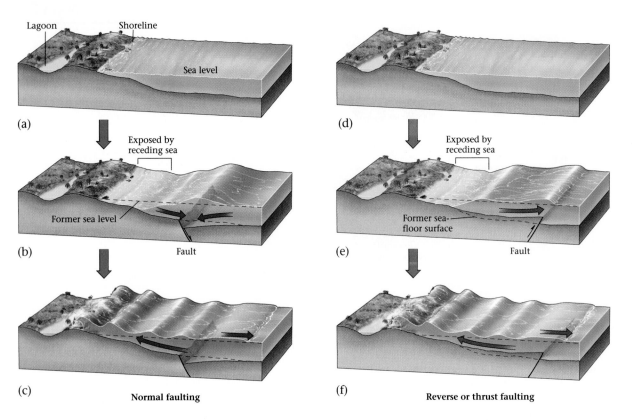

FIGURE 10.35 (a) Before a tsunami forms. (b) A normal fault creates a void, and water rushes to fill it. (c) The resulting waves move toward shore, where they build into a huge breaker. (d) A similar process happens in response to a reverse or thrust fault. (e) This time, the rising sea floor shoves up the water surface. (f) Large breakers develop.

NOTES

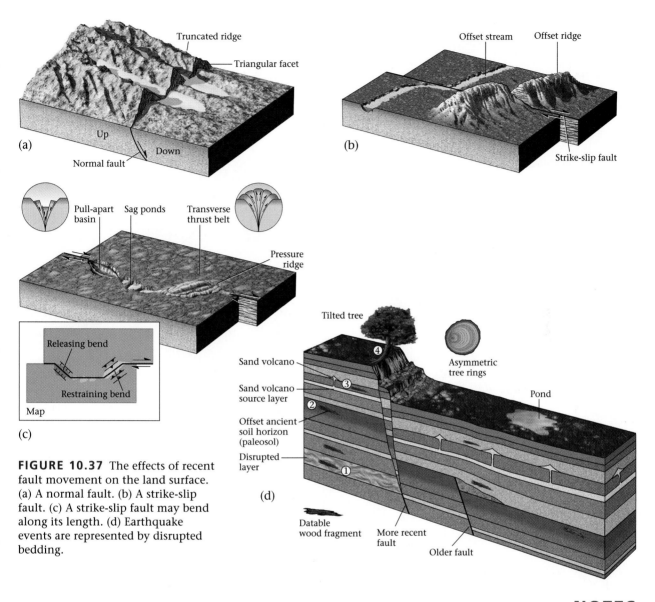

FIGURE 10.37 The effects of recent fault movement on the land surface. (a) A normal fault. (b) A strike-slip fault. (c) A strike-slip fault may bend along its length. (d) Earthquake events are represented by disrupted bedding.

NOTES

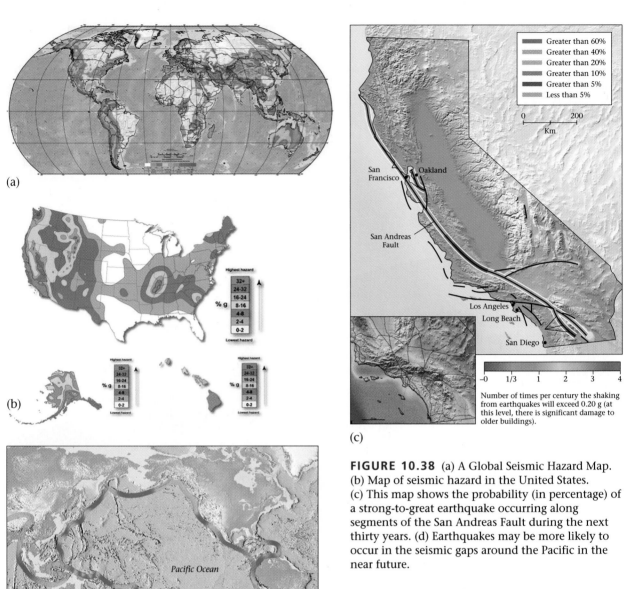

FIGURE 10.38 (a) A Global Seismic Hazard Map. (b) Map of seismic hazard in the United States. (c) This map shows the probability (in percentage) of a strong-to-great earthquake occurring along segments of the San Andreas Fault during the next thirty years. (d) Earthquakes may be more likely to occur in the seismic gaps around the Pacific in the near future.

NOTES

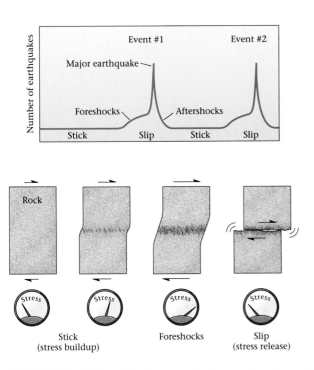

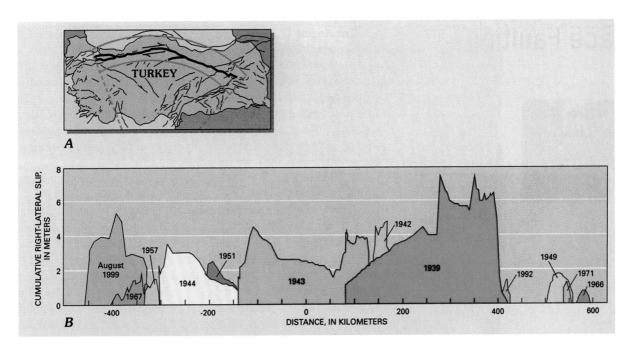

FIGURE 10.39 A sudden increase in the number of small earthquakes along a fault segment may be a possible precursor to a large earthquake along that segment.

FIGURE 10.40 (a) A map of Turkey, showing the Anatolian fault. (b) A graph representing regions that slipped during various earthquakes.

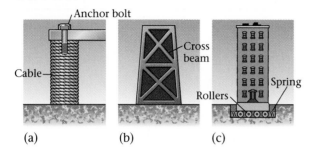

FIGURE 10.41 How to prevent damage and injury during an earthquake.

FIGURE 10.42 If an earthquake strikes, take cover under a sturdy table near a wall.

INTERLUDE C *Seeing Inside the Earth*

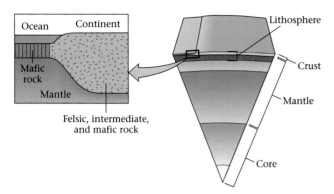

FIGURE C.1 The nineteenth-century three-layer image of the Earth, showing the crust, mantle, and core.

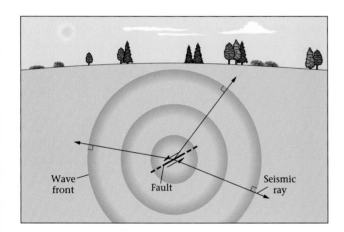

FIGURE C.2 An earthquake sends out waves in all directions.

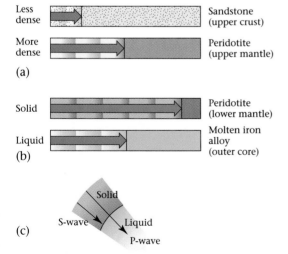

FIGURE C.3 (a) Seismic waves travel at different velocities in different rock types. (b) Seismic waves travel faster in solid peridotite than in a liquid like molten iron alloy. (c) Both P-waves and S-waves can travel through a solid, but only P-waves can travel through a liquid.

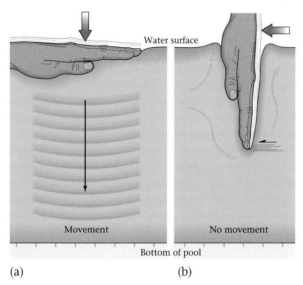

(a)   (b)

FIGURE C.4 (a) Pushing down on a liquid creates a compressive pulse (P-wave) that can travel through a liquid. (b) Shearing your hand through water does not generate a shear wave; the moving water simply flows past the water deeper down.

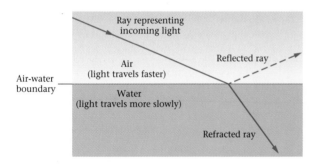

(a)

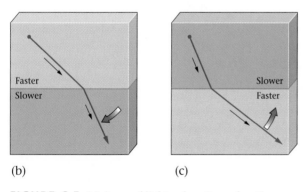

(b)   (c)

FIGURE C.5 (a) A ray of light, when it reaches the boundary between water and air, partly reflects and partly refracts. (b) A ray that enters a slower medium bends away from the boundary (like light reaching water from air). (c) A ray that enters a faster medium bends toward the boundary.

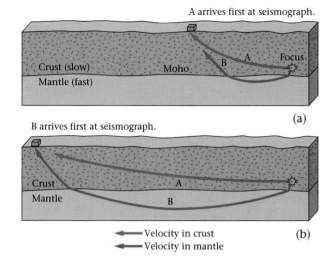

A arrives first at seismograph.

Crust (slow) Moho Mantle (fast) A B Focus

(a)

B arrives first at seismograph.

Crust Mantle A B

Velocity in crust
Velocity in mantle

(b)

FIGURE C.6 (a) At a nearby seismograph station, seismic waves traveling through the crust reach the seismograph first. (b) At a distant station, seismic waves traveling through the mantle reach the seismograph first.

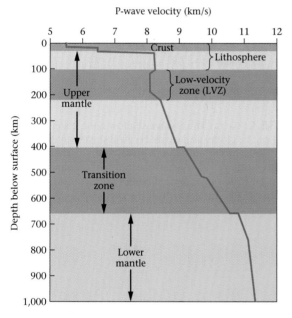

P-wave velocity (km/s)

Crust Lithosphere
Low-velocity zone (LVZ)
Upper mantle
Transition zone
Lower mantle

Depth below surface (km)

FIGURE C.7 The velocity of P-waves in the mantle changes with depth.

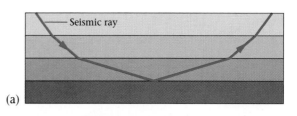

Seismic ray

(a)

(b)

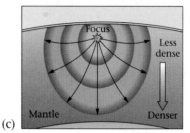

Focus Less dense
Mantle Denser

Curved rays in a mantle whose density increases gradually with depth

(c)

FIGURE C.8 (a) In a stack of layers in which seismic waves travel at different velocities (fastest in the lowest layer), a seismic ray gradually bends around and heads back to the surface. (b) If the mantle's density increased gradually with depth, the ray would define a smooth curve. (c) Since the velocity of seismic waves increases with depth, wave fronts are oblong and seismic rays curve.

NOTES

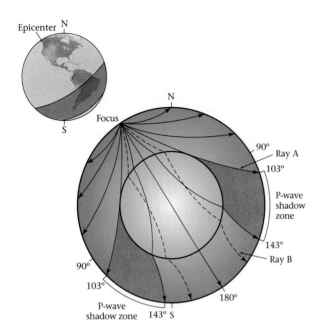

FIGURE C.9 P-waves do not arrive in the interval between 103° and 143° from an earthquake's epicenter, defining the P-wave shadow zone.

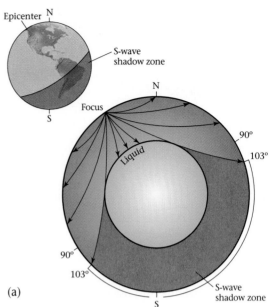

(a)

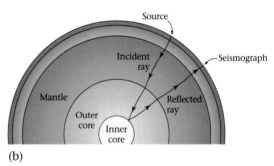

(b)

FIGURE C.10 (a) The S-wave shadow zone covers about a third of the globe, and exists because shear waves cannot pass through the liquid outer core. (b) The solid inner core was detected when seismologists observed that some seismic waves generated by nuclear explosions reflected off a boundary within the core.

NOTES

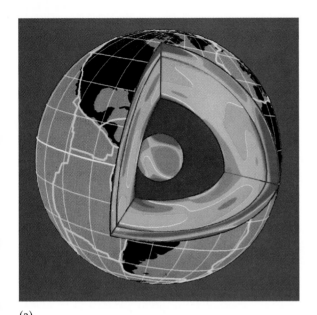

(a)

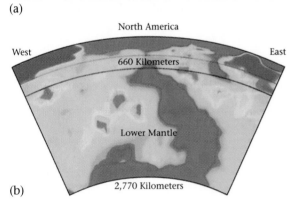

(b)

FIGURE C.11 Tomographic images show regions of faster and slower velocities.

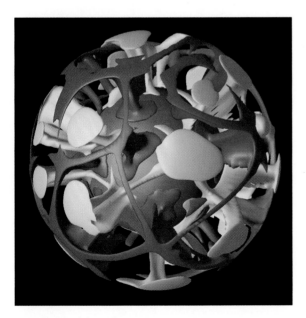

FIGURE C.12 This image is a computer-generated 3-D model depicting convective flow in the mantle.

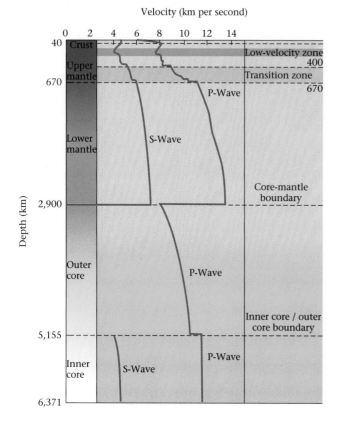

FIGURE C.14 The velocity-versus-depth profile for the Earth.

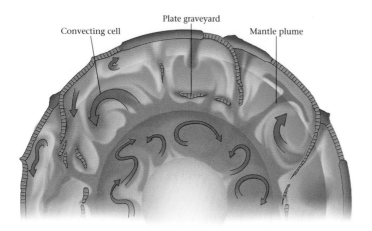

FIGURE C.16 The modern view of a complex and dynamic Earth interior.

CHAPTER 11 *Crags, Cracks, and Crumples: Crustal Deformation and Mountain Building*

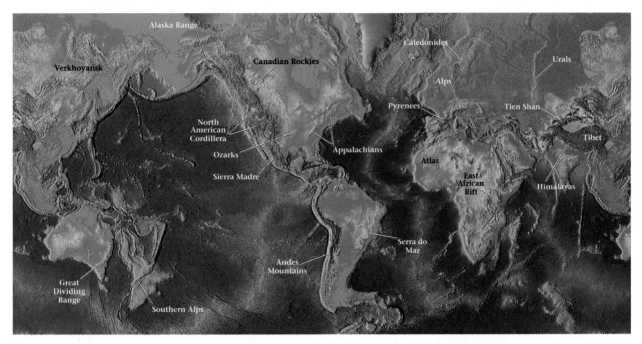

FIGURE 11.2 Digital map of world topography, showing the locations of major mountain ranges.

NOTES

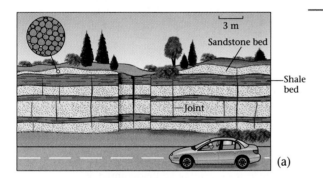

(a)

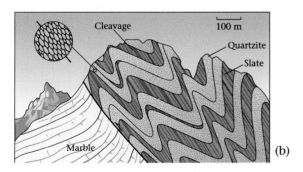

(b)

FIGURE 11.3 (a) This road cut exposes flat-lying beds of Paleozoic shale and sandstone along a highway. (b) Diagram of an Alpine outcrop.

Earth: Portrait of a Planet
W. W. Norton & Company, Inc., 500 Fifth Avenue, New York, N.Y. 10110

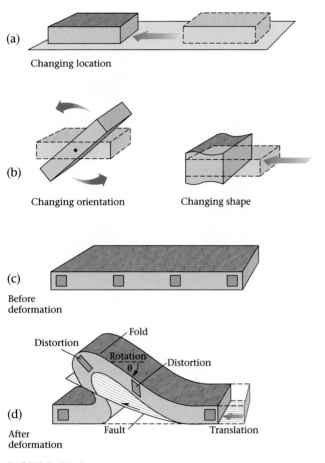

FIGURE 11.4 The components of deformation. (a) Change of location. (b) Change of orientation. (c) Change of shape. (d) Folds and faults represent deformation.

NOTES

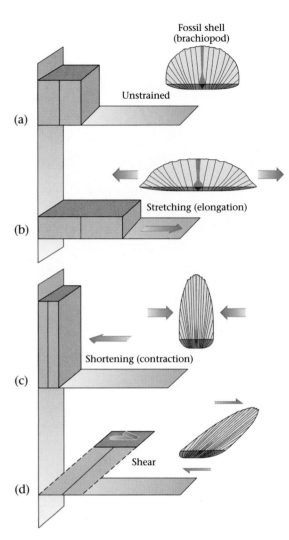

FIGURE 11.6 Different kinds of strain. (a) An unstrained cube and an unstrained fossil shell. (b) Stretching changes the cube into a brick. (c) Shortening changes the cube into a brick whose long dimension lies perpendicular to the shortening direction. (d) Shear strain tilts the cube over and transforms it into a parallelogram.

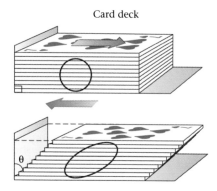

FIGURE 11.7 Simulation of shear strain using a deck of cards.

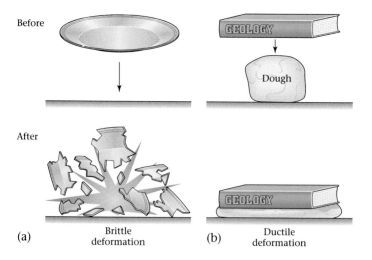

FIGURE 11.8 (a) Brittle deformation—drop a plate and it shatters. (b) Ductile deformation—squash a soft ball of dough beneath a book.

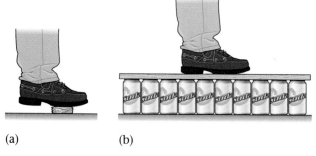

FIGURE 11.9 (a) Stand on a single can—apply enough force to the can to crush it. (b) Stand on a board resting on 100 cans—it does not crumple.

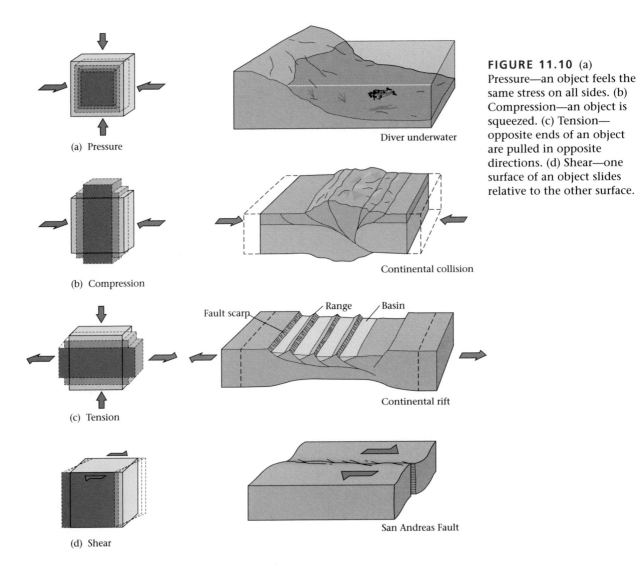

(a) Pressure

Diver underwater

(b) Compression

Continental collision

Fault scarp   Range   Basin

(c) Tension

Continental rift

(d) Shear

San Andreas Fault

FIGURE 11.10 (a) Pressure—an object feels the same stress on all sides. (b) Compression—an object is squeezed. (c) Tension—opposite ends of an object are pulled in opposite directions. (d) Shear—one surface of an object slides relative to the other surface.

NOTES

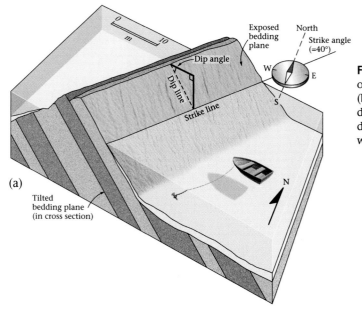

(a)

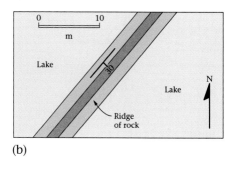

(b)

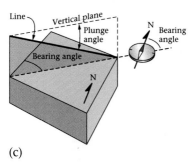

(c)

FIGURE 11.11 (a) Strike and dip—measures the orientation of planar structures like these tilted beds. (b) On a map, the line segment represents the strike direction, while the tick on the segment represents the dip direction. (c) To specify the orientation of a line, we use plunge and bearing.

NOTES

(a)

(b)

(c)

(d)

FIGURE 11.13 (a) The San Andreas Fault displacing a creek. (b) What a geologist sees looking down on the San Andreas fault. (c) A road cut in the Rocky Mountains, showing a fault offsetting strata in cross section. (d) What a geologist sees looking at the Rocky Mountain road cut.

NOTES

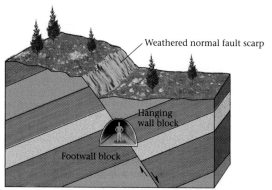

(a)

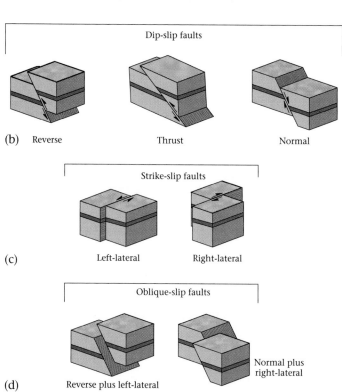

(b)   Reverse   Thrust   Normal

(c)   Left-lateral   Right-lateral

(d)   Reverse plus left-lateral   Normal plus right-lateral

FIGURE 11.14 (a) A hanging-wall block and footwall block. (b) Three types of dip-slip faults. (c) Two types of strike-slip faults. (d) Two examples of oblique-slip faults.

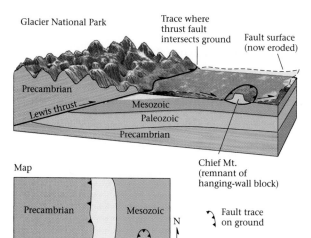

FIGURE 11.15 This large thrust fault (the Lewis thrust) puts older rock (Precambrian) over younger rock (Mesozoic). On the geologic map of the region, the triangular barbs point to the hanging-wall block.

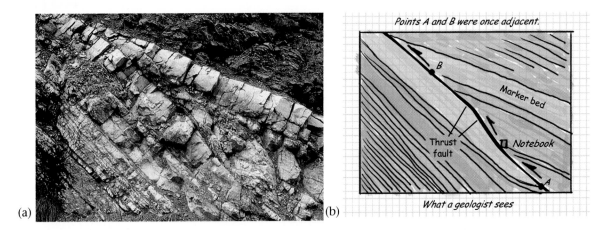

FIGURE 11.16 (a) A thrust fault—a distinct layer has been offset. (b) A geologist's sketch emphasizes the offset.

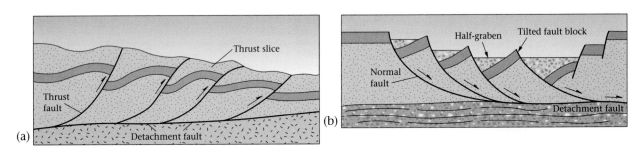

FIGURE 11.18 (a) In this thrust-fault system, several related thrust faults merge at depth with a detachment fault. (b) A normal-fault system consists of several related normal faults.

FIGURE 11.19 (a) Horsts and grabens cutting through marble. (b) A geologist's sketch of the quarry wall.

NOTES

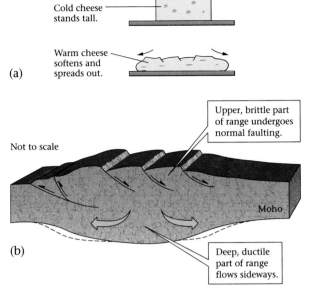

FIGURE 11.32 (a) Cheese spreads out sideways as it warms up and softens. (b) Similarly, mountain belts spread out sideways once they reach a certain thickness.

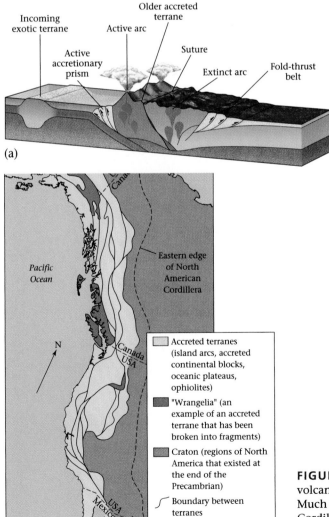

FIGURE 11.33 (a) In a convergent-margin orogen, volcanic arcs form, and there may be compression. (b) Much of the western portion of the North American Cordillera consists of accreted terranes.

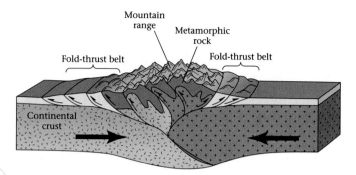

FIGURE 11.34 In a collisional orogen, two continents collide.

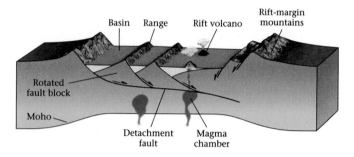

FIGURE 11.35 When the crust stretches in a continental rift, rift-related mountains form, as do normal faults.

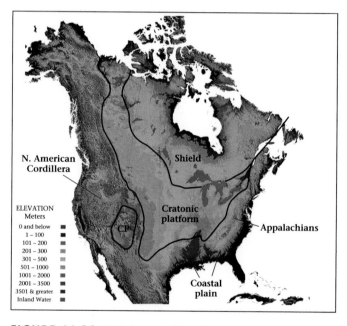

FIGURE 11.36 Digital map of North America.

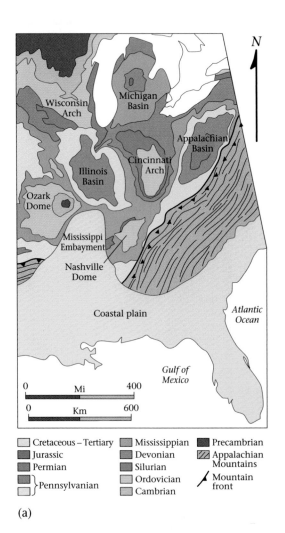

(a)

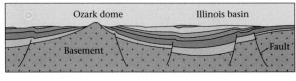

(b)

FIGURE 11.37 (a) Geologic map of the mid-continent region of the United States, showing the basins and domes and the faults that cut the region. (b) This cross section illustrates the geometry of the regional basins and domes.

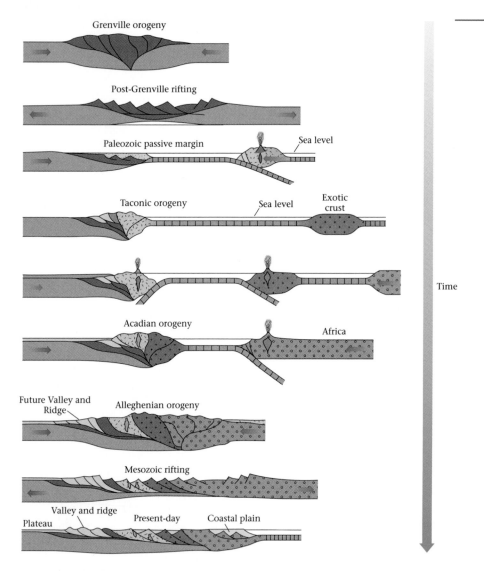

Grenville orogeny

Post-Grenville rifting

Paleozoic passive margin — Sea level

Taconic orogeny — Sea level — Exotic crust

Acadian orogeny — Africa

Future Valley and Ridge — Alleghenian orogeny

Mesozoic rifting

Plateau — Valley and ridge — Present-day — Coastal plain

Time

FIGURE 11.38 These idealized stages show the tectonic evolution of the Appalachian Mountains.

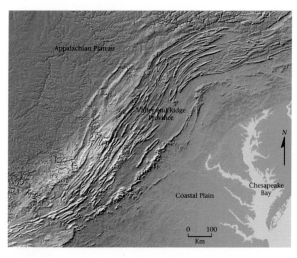

FIGURE 11.39 Relief map of the Valley and Ridge Province, Pennsylvania to Virginia. The ridges, which outline the shapes of plunging folds in the fold-thrust belt, are cmposed of resistant sandstone beds.

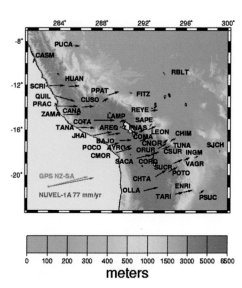

FIGURE 11.40 Convergence between the Nazca Plate and South America creates the broad Andean orogen.

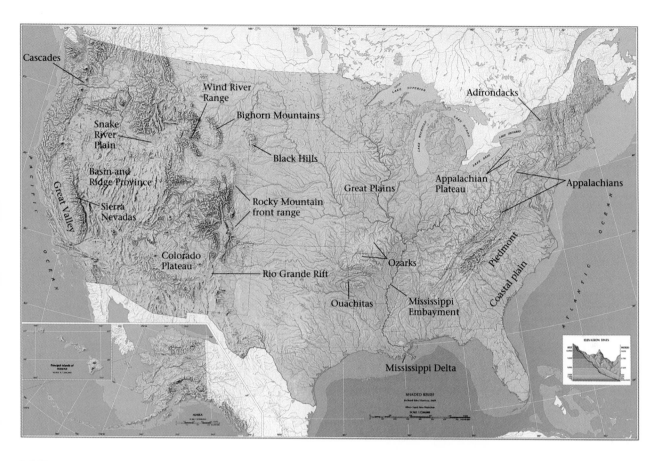

REFERENCE MAP Topography of the United States.

NOTES

INTERLUDE D   *Memories of Past Life: Fossils and Evolution*

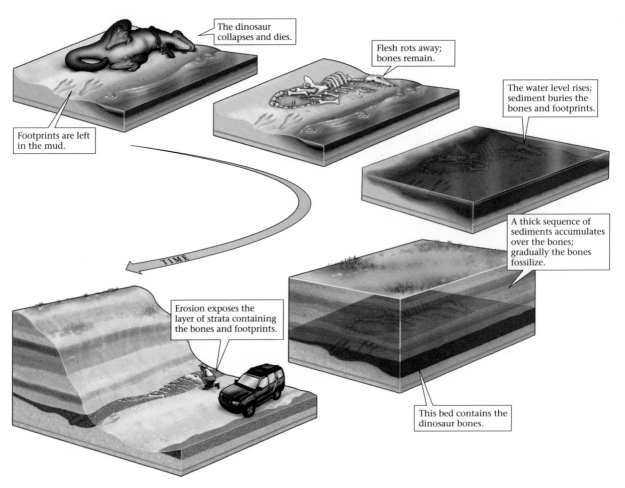

FIGURE D.4 How a dinosaur eventually becomes a fossil.

NOTES

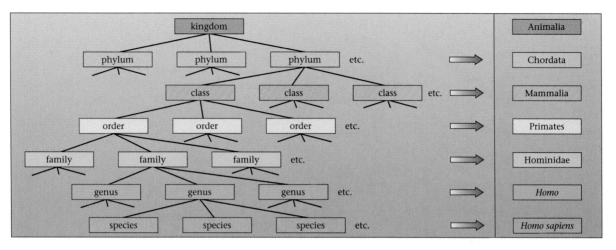

FIGURE D.8 The taxonomic subdivisions.

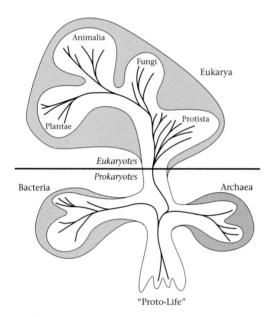

FIGURE D.9 The basic kingdoms of life on Earth.

NOTES

FIGURE D.10 Examples of the diversity of gastropods (snails).

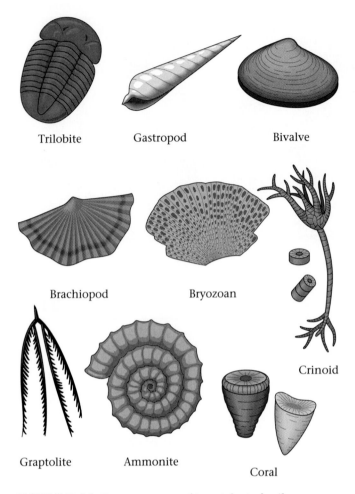

FIGURE D.11 Common types of invertebrate fossils.

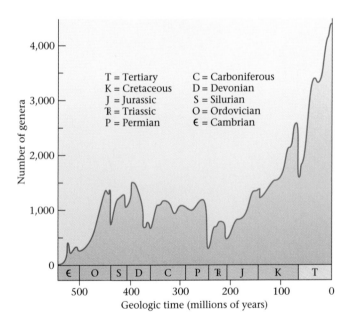

FIGURE D.12 This graph illustrates the variation in diversity of life with time.

CHAPTER 12 *Deep Time: How Old Is Old?*

FIGURE 12.3 The difference between relative and numerical age.

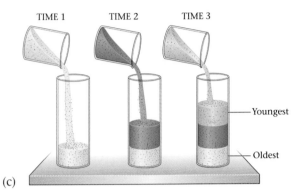

(c)

FIGURE 12.4 Geologic principles.
(c) Superposition. (f, g) Original
continuity. (h) Cross-cutting relations.
(i) Inclusions. (j) Baked contacts.

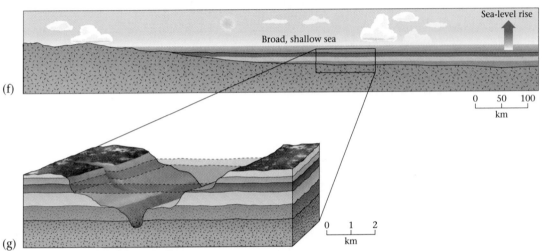

(f)

(g)

NOTES

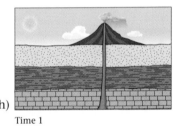

(h)

Time 1   Time 2

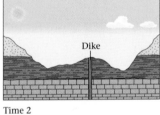

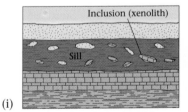

(i)

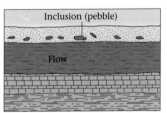

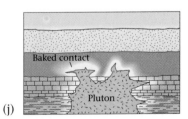

(j)

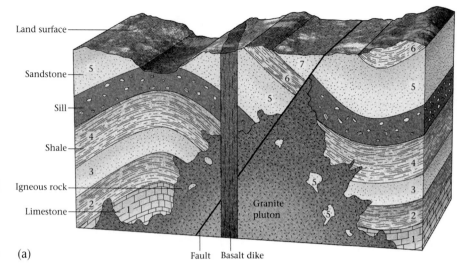

Land surface

Sandstone

Sill

Shale

Igneous rock

Limestone

Granite
pluton

Fault   Basalt dike

(a)

FIGURE 12.5 (a) Geologic
principles allow us to
interpret the sequence of
events leading to the
development of the features
shown here. (b) Sequence of
geologic events leading to the
geology shown in (a).

NOTES

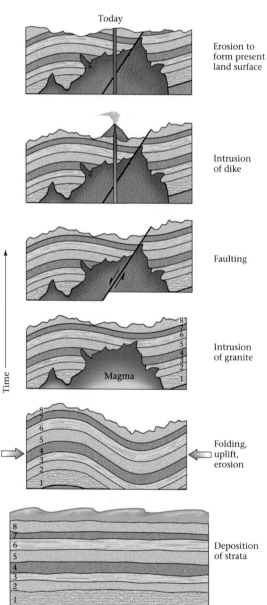

Today

Erosion to
form present
land surface

Intrusion
of dike

Faulting

Time

Magma

Intrusion
of granite

Folding,
uplift,
erosion

Deposition
of strata

Oldest

(b)

Range

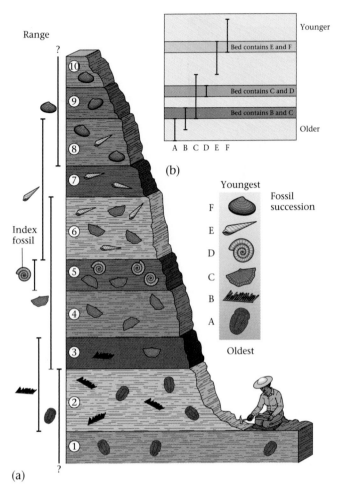

Index
fossil

(a)

(a)

(b)

Youngest

Fossil
succession

F
E
D
C
B
A

Oldest

FIGURE 12.7 (a) The principle of fossil succession.
(b) Overlapping fossil ranges can be used to limit the relative age
of a given bed and to determine the relative ages of beds.

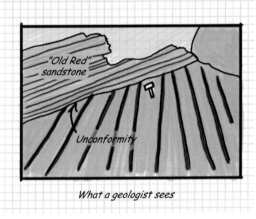

(b)

FIGURE 12.8 (a) The Siccar Point
unconformity in Scotland. (b) A geological
interpretation of the unconformity.

NOTES

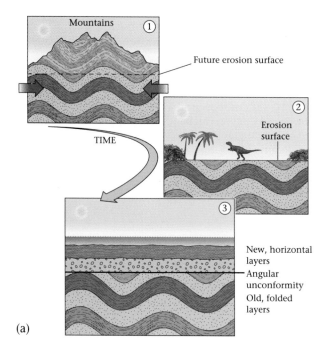

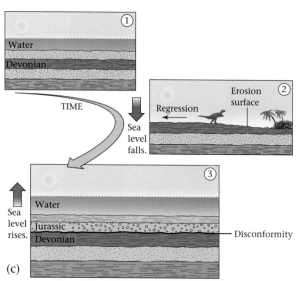

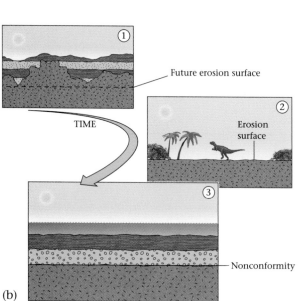

FIGURE 12.9 (a) The stages during the development of an angular unconformity. (b) The stages during the development of a nonconformity. (c) The stages during the development of a disconformity.

NOTES

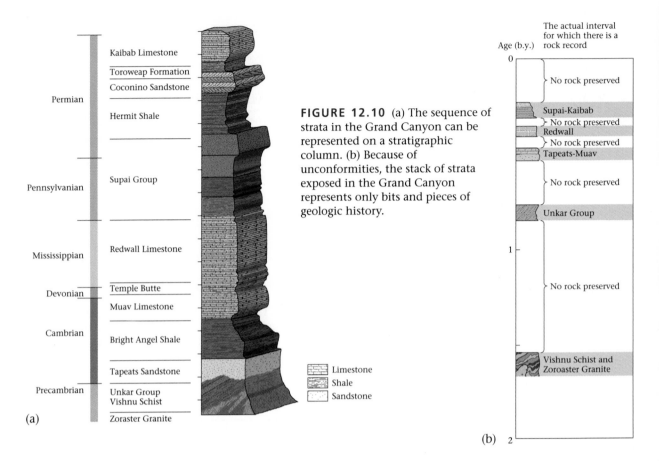

FIGURE 12.10 (a) The sequence of strata in the Grand Canyon can be represented on a stratigraphic column. (b) Because of unconformities, the stack of strata exposed in the Grand Canyon represents only bits and pieces of geologic history.

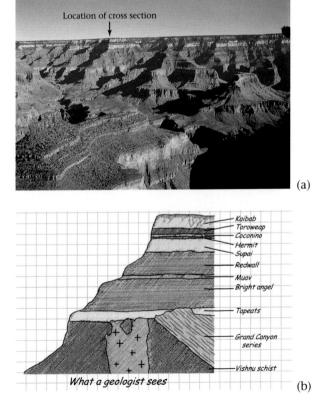

FIGURE 12.11 (a, b) The succession of rocks in the Grand Canyon can be divided into formations.

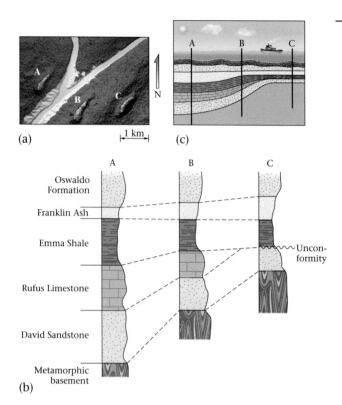

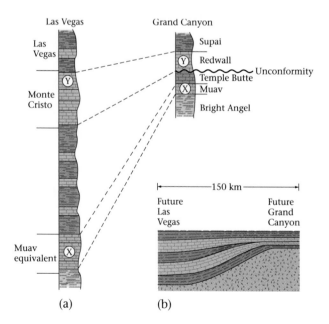

FIGURE 12.12 The principle of lithologic correlation. (a) These three outcrops of rock are a couple of kilometers from each other. (b) The stratigraphic sections at each location are somewhat different, but the columns can be correlated with one another by matching rock types. (c) Geologists reconstruct a sedimentary basin using correlation.

FIGURE 12.13 The principle of fossil correlation: (a) Because they both contain "Y-age" fossils, we can say that the Redwall Limestone of the Grand Canyon correlates with the Monte Cristo Limestone near Las Vegas. (b) In the Paleozoic era, a sedimentary basin thinned radically between Las Vegas and the Grand Canyon.

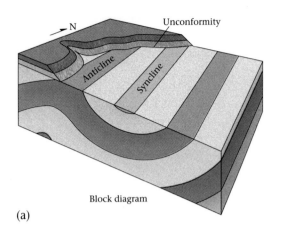

Block diagram

(a)

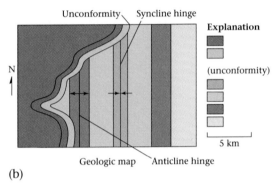

Unconformity   Syncline hinge

Explanation

(unconformity)

N

5 km

Geologic map   Anticline hinge

(b)

FIGURE 12.14 (a) An angular unconformity with horizontal strata above and folded strata below. (b) The geologic map shows what this ground surface would look like if viewed from above.

FIGURE 12.15 (a) The geologic column was constructed by determining the relative ages of stratigraphic columns from around the world. (b) By correlation, the strata in the columns can be stacked in a sequence representing most of geologic time.

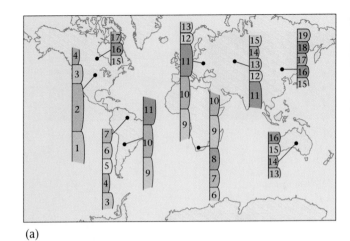

(a)

Eon	Era	Period	Epoch
Phanerozoic	Cenozoic	Quaternary	Holocene
			Pleistocene
		Tertiary	Pliocene
			Miocene
			Oligocene
			Eocene
			Paleocene
	Mesozoic	Cretaceous	
		Jurassic	
		Triassic	
	Paleozoic	Permian	Pennsylvanian
		Carboniferous	Mississippian
		Devonian	
		Silurian	
		Ordovician	
		Cambrian	
Precambrian	Proterozoic		
	Archean		

(b)   Geologic Column

NOTES

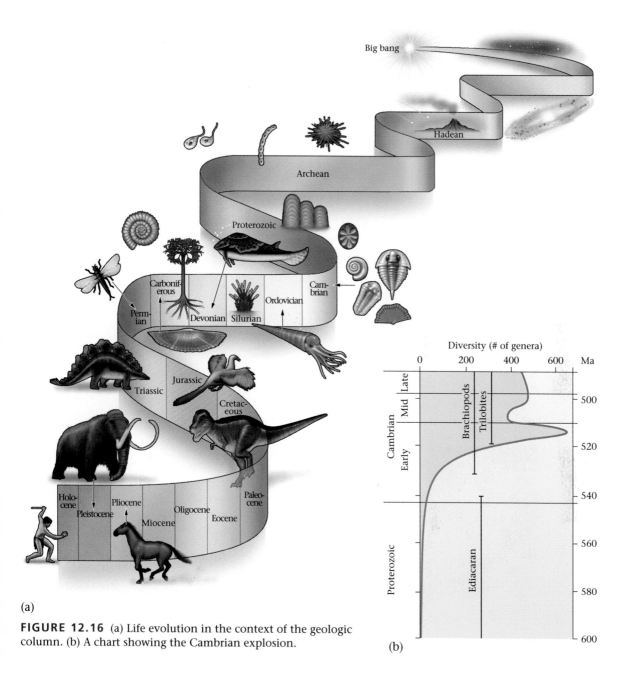

FIGURE 12.16 (a) Life evolution in the context of the geologic column. (b) A chart showing the Cambrian explosion.

NOTES

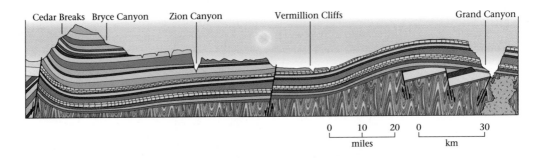

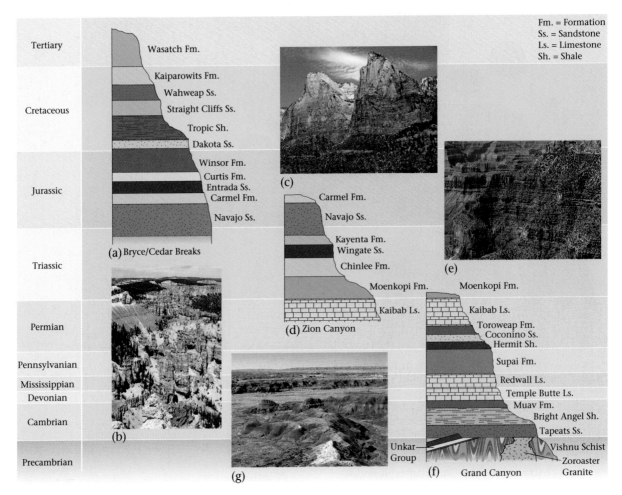

FIGURE 12.17 The correlation of strata among the various national parks of Arizona and Utah: (a, b) Bryce Canyon/Cedar Breaks, (c, d) Zion Canyon, (e, f) the Grand Canyon, (g) the Painted Desert. The inset at the top shows a cross section of the region.

NOTES

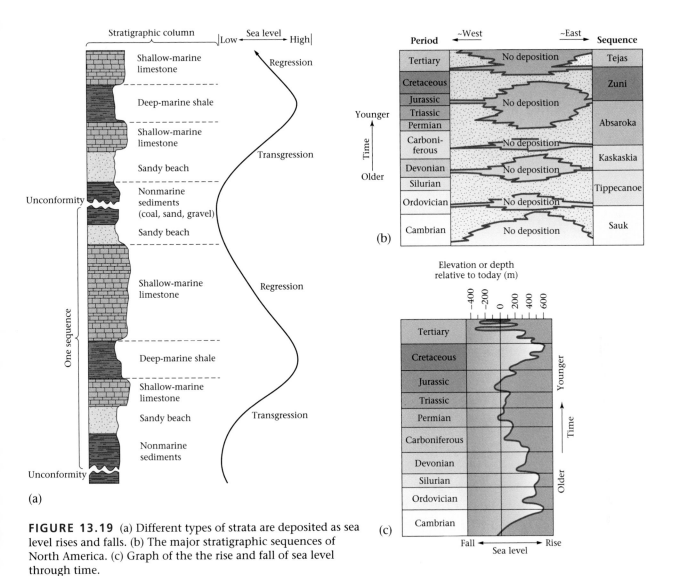

FIGURE 13.19 (a) Different types of strata are deposited as sea level rises and falls. (b) The major stratigraphic sequences of North America. (c) Graph of the the rise and fall of sea level through time.

NOTES

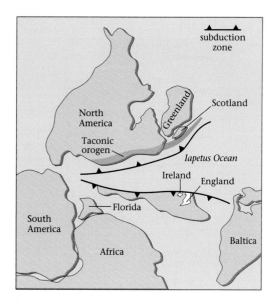

FIGURE 13.20 This paleogeographical map shows the movement of the Avalon microcontinent toward its ultimate collision with the eastern margin of North America.

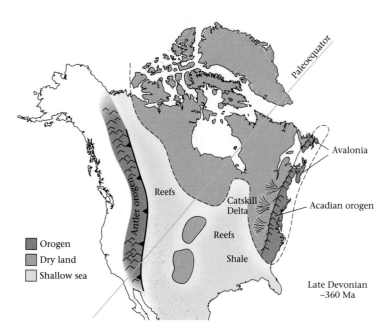

FIGURE 13.21 This paleogeographical map of North America depicts the distribution of land and sea and the position of mountain belts in the Late Devonian Period.

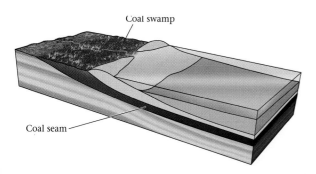

FIGURE 14.17 Sea level transgresses and regresses over time, with the result that a coal swamp along the coast migrates inland and the swamp's deposits eventually get buried by other strata.

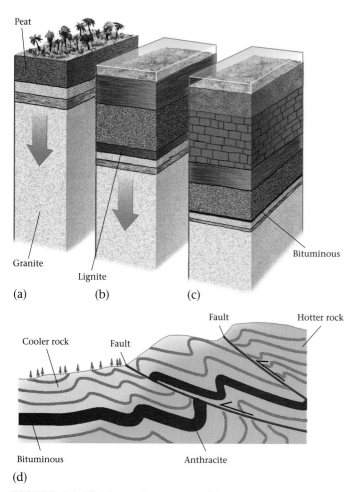

FIGURE 14.18 The evolution of coal from peat.

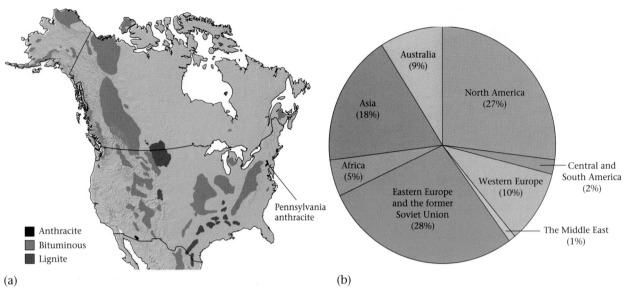

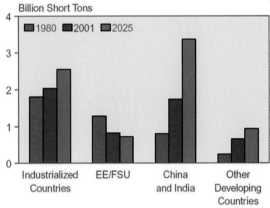

Anthracite
Bituminous
Lignite

Pennsylvania anthracite

(a)

(b)

Australia (9%)

North America (27%)

Asia (18%)

Africa (5%)

Eastern Europe and the former Soviet Union (28%)

Western Europe (10%)

Central and South America (2%)

The Middle East (1%)

Billion Short Tons

1980 2001 2025

Industrialized Countries

EE/FSU

China and India

Other Developing Countries

(c) **World Coal Consumption by Region**

FIGURE 14.20 (a) The distribution of coal reserves in North America. (b) The distribution of coal reserves by region. (c) World coal consumption by region.

NOTES

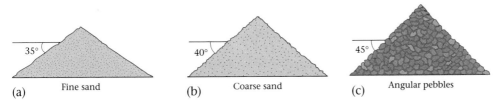

(a)   35°   Fine sand

(b)   40°   Coarse sand

(c)   45°   Angular pebbles

FIGURE 16.14 The angle of repose is the steepest slope that a pile of unconsolidated sediment can have and remain stable.

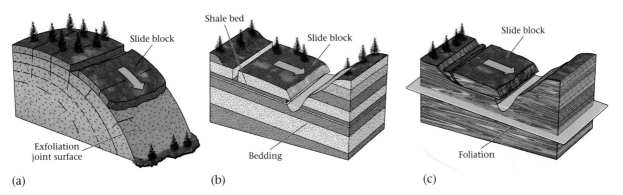

(a)   (b)   (c)

Slide block — Exfoliation joint surface — Shale bed — Slide block — Bedding — Slide block — Foliation

FIGURE 16.15 Different kinds of surfaces become failure surfaces in different geologic settings.

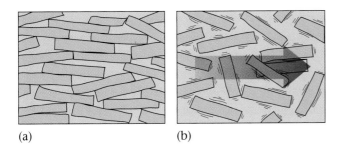

(a)   (b)

FIGURE 16.16 (a) In a quick clay, before shaking, the grains stick together. (b) During shaking, the grains become suspended in water, and the formerly solid mass becomes a movable slurry.

NOTES

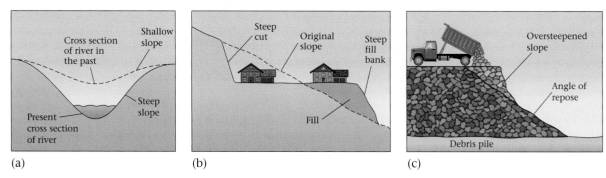

FIGURE 16.17 Slope angles may become steeper, making the slopes unstable.

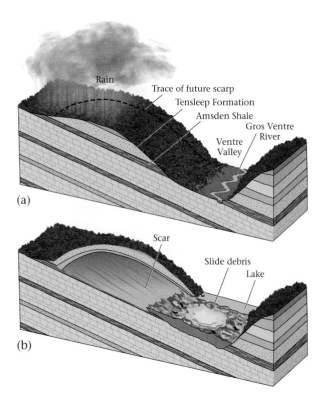

FIGURE 16.18 (a) The huge Gros Ventre slide took place after heavy rains had seeped into the ground. (b) After the slide moved, it filled the river valley and dammed the river, creating a lake.

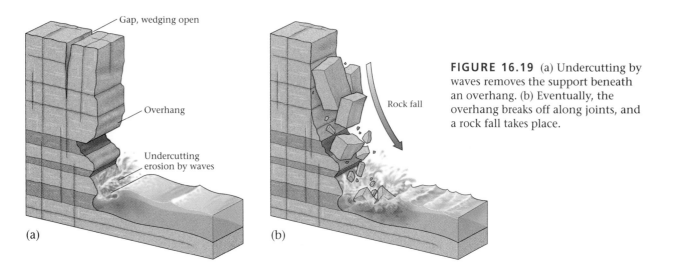

Gap, wedging open

Overhang

Undercutting erosion by waves

(a)

Rock fall

(b)

FIGURE 16.19 (a) Undercutting by waves removes the support beneath an overhang. (b) Eventually, the overhang breaks off along joints, and a rock fall takes place.

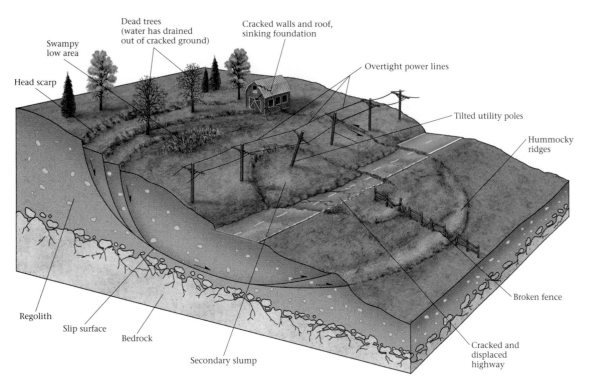

Swampy low area

Dead trees (water has drained out of cracked ground)

Cracked walls and roof, sinking foundation

Overtight power lines

Head scarp

Tilted utility poles

Hummocky ridges

Regolith

Slip surface

Bedrock

Secondary slump

Broken fence

Cracked and displaced highway

FIGURE 16.22 The features shown here indicate that a large slump is beginning to develop.

NOTES

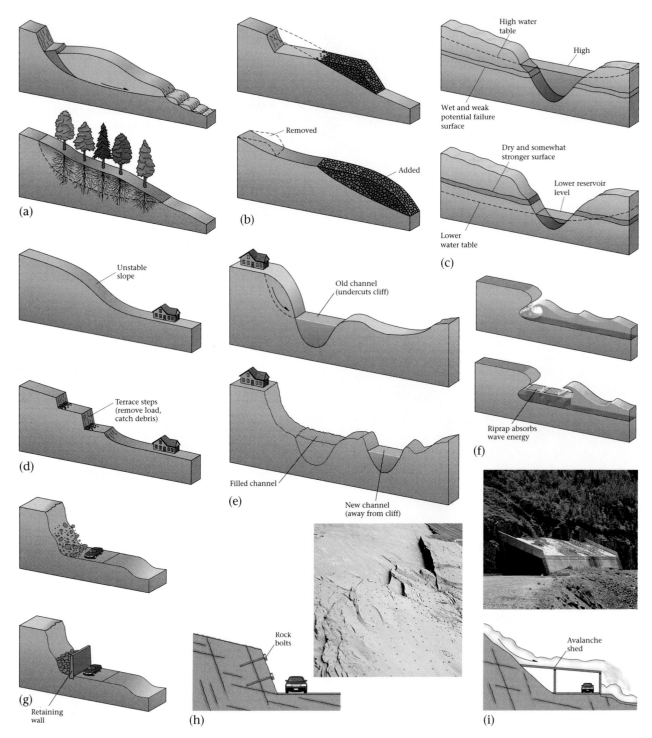

FIGURE 16.23 A variety of remedial steps can stabilize unstable ground.

NOTES

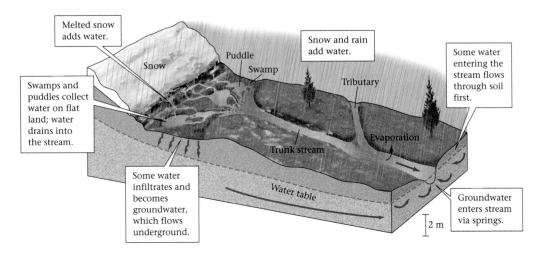

FIGURE 17.2 Excess surface water comes from rain, melting ice or snow, and groundwater springs.

NOTES

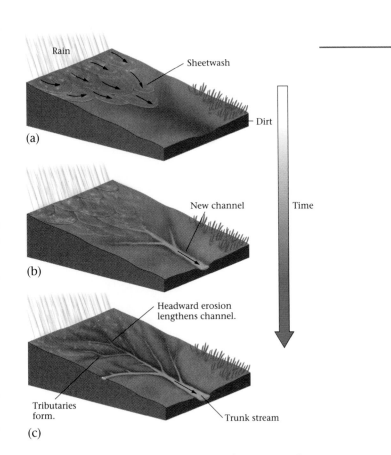

FIGURE 17.3 (a) Drainage on a slope first occurs when sheetwash, overlapping films or sheets of water, moves downslope. (b) Where the sheetwash happens to move a little faster, it scours a channel. (c) The channel grows upslope, a process called headward erosion, and new tributary channels form.

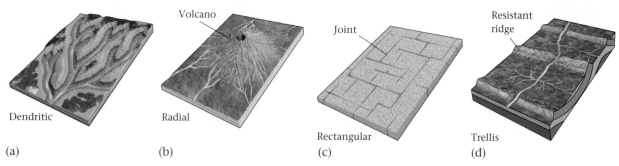

FIGURE 17.4 Patterns of drainage networks.

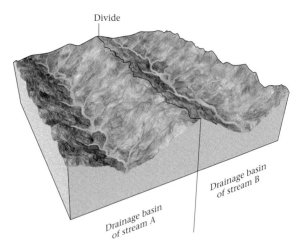

FIGURE 17.5 A drainage divide is a relatively high ridge that separates one drainage basin from another.

FIGURE 17.6 The Mississippi drainage basin is one of several drainage basins in North America.

NOTES

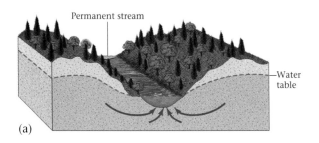

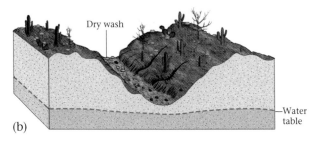

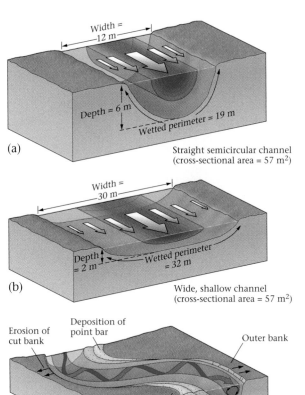

FIGURE 17.7 (a) Permanent stream—below the water table. (b) Dry wash—above the water table.

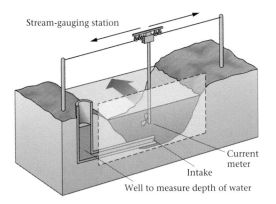

FIGURE 17.8 Measuring a stream's discharge at a stream-gauging station.

FIGURE 17.9 (a) In a straight semicircular channel, the maximum velocity occurs near the surface in the center of the stream. (b) The maximum velocity also occurs in the center of a wide, shallow channel. (c) In a curved channel, the fastest flow shifts toward the outer edge of the stream, over the thalweg.

NOTES

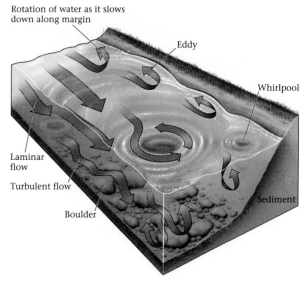

Rotation of water as it slows
down along margin

Eddy

Whirlpool

Laminar
flow

Turbulent flow

Boulder

Sediment

FIGURE 17.10 In a turbulent flow the water swirls in curving paths and becomes caught in eddies.

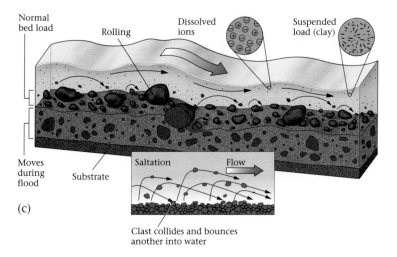

Normal
bed load

Rolling

Dissolved
ions

Suspended
load (clay)

Moves
during
flood

Substrate

Saltation

Flow

(c)

Clast collides and bounces
another into water

FIGURE 17.11 (c) Streams transport sediment in many forms.

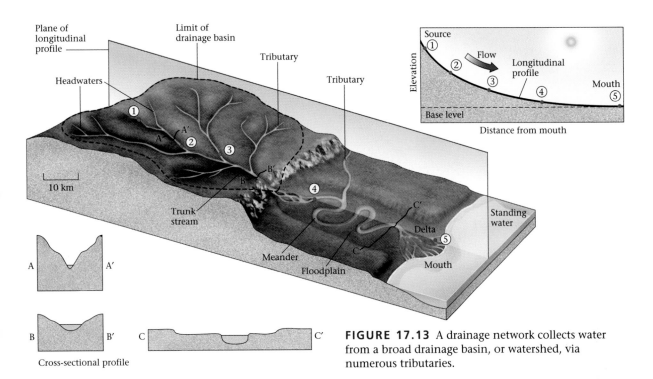

FIGURE 17.13 A drainage network collects water from a broad drainage basin, or watershed, via numerous tributaries.

NOTES

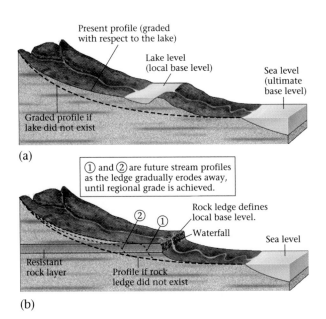

(a)

(b)

FIGURE 17.14 (a) A lake acts as a local base level. (b) A resistant rock ledge also acts as a local base level.

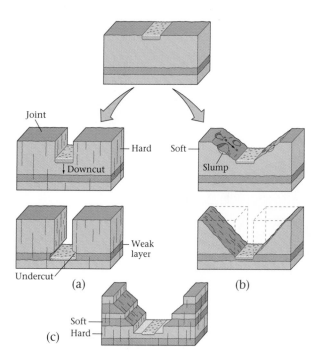

FIGURE 17.16 (a) If downcutting by a stream happens faster than mass wasting alongside the stream, a slot canyon forms. (b) If mass wasting takes place as fast as downcutting occurs, a V-shaped valley develops. (c) In regions where the stream downcuts through alternating hard and soft layers, a stair-step canyon forms.

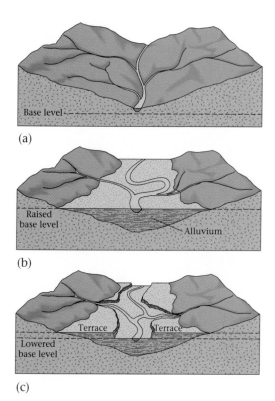

FIGURE 17.17 The evolution of alluvium-filled valleys.

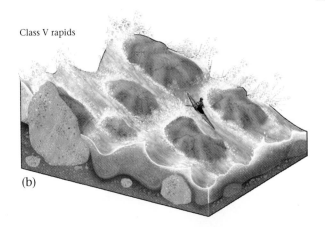

Class V rapids

(b)

FIGURE 17.18 (b) Classe V rapids should only be navigated by experts!

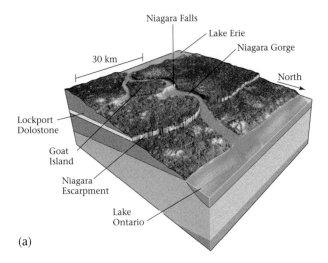

Niagara Falls

Lake Erie

Niagara Gorge

30 km

North

Lockport
Dolostone

Goat
Island

Niagara
Escarpment

Lake
Ontario

(a)

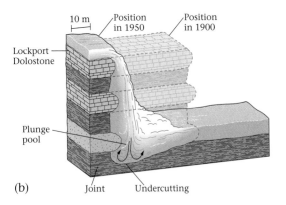

10 m

Position
in 1950

Position
in 1900

Lockport
Dolostone

Plunge
pool

(b)

Joint   Undercutting

FIGURE 17.20 (a) Niagara Falls exists because Lake Erie lies at a higher elevation than Lake Ontario. (b) This cross section of the falls shows how undercutting of the soft shale layers eventually causes the resistant layers of dolostone to break off at joints.

Time 1

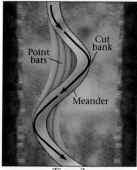

Time 2

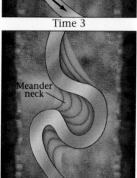

Time 3

Time 4

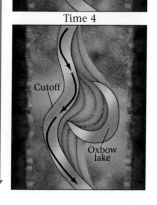

Time

(b)

(c)

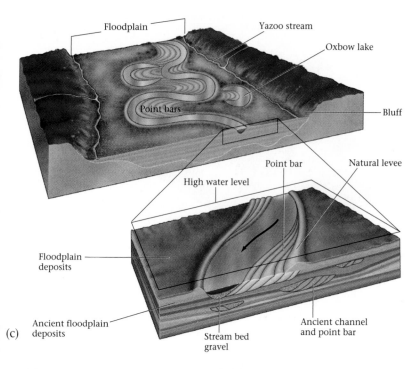

FIGURE 17.22 (b) Erosion occurs faster on the outer bank of a stream's curve, while deposition takes place on the inner curve. (c) The landforms of a meandering stream.

NOTES

Vulnerable Site for a Metropolis
Squeezed between Lake Pontchartrain and the Mississippi River, New Orleans is surrounded by water that could inundate it in a strong hurricane or river flood. Much of it sits below sea level in what geologists call "the bowl."

MAP ELEVATIONS
This view has a twentyfold vertical exaggeration to show changes in terrain:

35 FEET ABOVE SEA LEVEL

Red areas are within 1 foot of sea level or lower.

SEA LEVEL
6 FEET BELOW SEA LEVEL

The Worst-Case Hurricane
Sustained counterclockwise winds blowing from north to south could push water from Lake Pontchartrain over levees and into the northern sections of the city.

EVACUATION WORRIES There are three main routes out of the city, all of them problematic:

1 Interstate 10 is prone to flooding where it passes over a corner of the lake.

2 The 24-mile Lake Pontchartrain Causeway is closed when winds exceed 50 mph.

3 Interstates 10 and 59 would very likely be clogged with traffic from vulnerable areas on the Mississippi Gulf coast.

Sources: Louisiana Water Resources Research Institute, Louisiana State University; U.S. Geological Survey; U.S. Army Corps of Engineers

CROSS SECTION OF THE BOWL This slice through the city corresponds to the white line below.

AVG. CREST 14 FT.
ST. LOUIS CATHEDRAL, FRENCH QUARTER
GENTILLY RIDGE
LAKE LEVEL IN MODERATE HURRICANE 11.5 14 FT.
RIVER SEA LEVEL
NORMAL LAKE LEVEL: 1 FT. ABOVE SEA LEVEL

Flooding Scenarios

EXTREMELY HEAVY RAINFALL Areas at sea level and below are flooded.

LAKE LEVEES OVERFLOW Most neighborhoods are swamped, except some closest to the riverbank. This view shows flooded areas whose elevation is at most 7 feet above sea level.

COMMUNITY HAVEN Some suggest walling off a section of the city in anticipation of a catastrophic flood, shown here at 10 feet. The haven would spare downtown, the French Quarter, and some neighborhoods while offering refuge for others.

DOWNTOWN
FRENCH QUARTER

William McNulty and Bill Marsh/The New York Times

FIGURE 17.23 New Orleans lies in a low area between the Mississippi River and Lake Pontchartrain.

NOTES

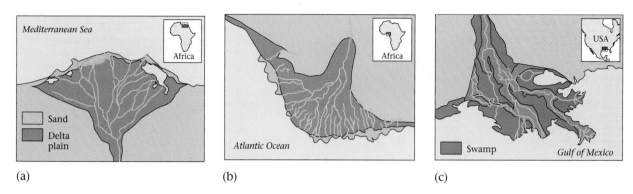

FIGURE 17.24 (a) The Nile is a Δ-shaped delta. (b) The Niger is an arc-like delta. (c) The Mississippi is a bird's-foot delta.

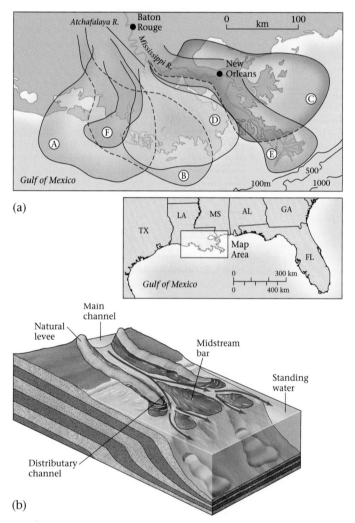

Delta deposit	Age (years)
F	400 b.p. – present
E	1,000 b.p. – present
D	2,500 b.p. – 800 b.p.
C	4,000 b.p. – 2,000 b.p.
B	5,500 b.p. – 3,800 b.p.
A	7,500 b.p. – 5,000 b.p.

NOTES

FIGURE 17.25 (a) The map shows the different, dated lobes of the Mississippi Delta and the different channels that served as their source. Inset shows a current view of the delta, relative to Louisiana. (b) When a stream enters standing water, it deposits more sediment in the center of the channel than along the margins because the formerly fast-moving water at the center carried more sediment.

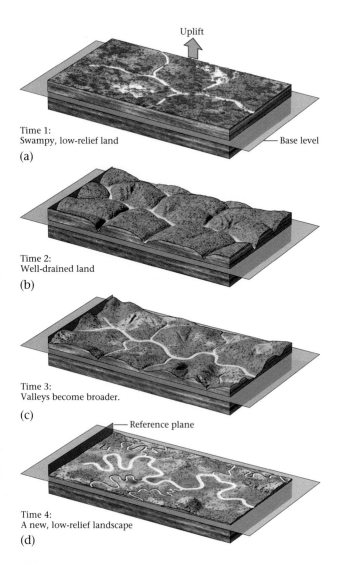

Uplift

Time 1:
Swampy, low-relief land
(a)

Base level

Time 2:
Well-drained land
(b)

Time 3:
Valleys become broader.
(c)

Reference plane

Time 4:
A new, low-relief landscape
(d)

FIGURE 17.26 (a) A fluvial landscape is first uplifted. (b) Then, the stream cuts down into the plain. (c) Later, the landscape consists of rounded hills dissected by tributaries. (d) Still later, only a few remnant hills are left.

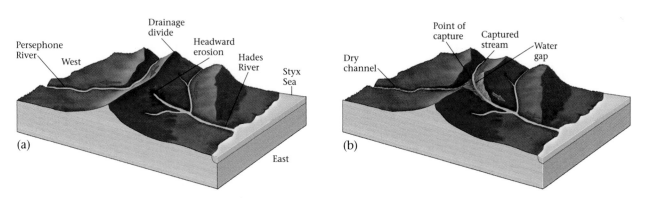

Persephone River
West

Drainage divide

Headward erosion

Hades River

Styx Sea

East

(a)

Point of capture

Captured stream

Water gap

Dry channel

(b)

FIGURE 17.27 (a) A drainage divide separates the Hades River drainage from the Persephone River drainage. (b) When the source of the Hades River reaches the channel of the Persephone River, Hades (the pirate stream) captures Persephone and carries off its water to the Styx Sea.

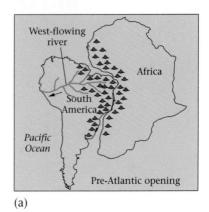

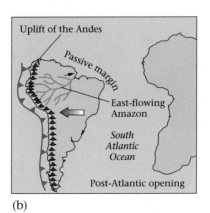

FIGURE 17.28 (a) Early Mesozoic Era. (b) Late Mesozoic Era.

(a)

(b)

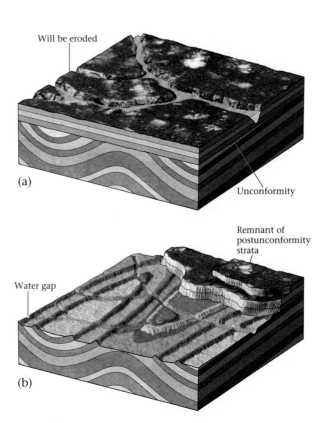

(a)

(b)

FIGURE 17.29 (a) A superposed stream first establishes its geometry while flowing over uniform, flat layers above an unconformity. (b) The stream gradually erodes away the layers and exposes underlying rock with a different structure.

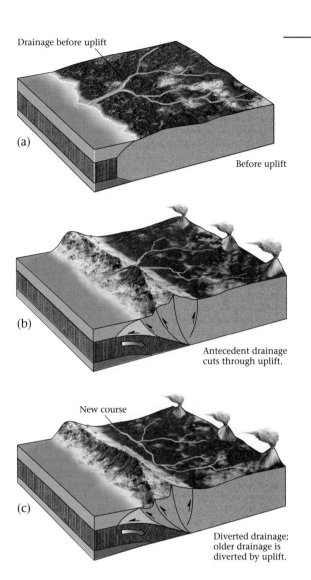

Drainage before uplift

(a)

Before uplift

(b)

Antecedent drainage
cuts through uplift.

New course

(c)

Diverted drainage;
older drainage is
diverted by uplift.

FIGURE 17.30 (a) An antecedent stream flows
across the land to the sea. (b) A mountain range
develops across the path of the stream. (c) If uplift
happens faster than erosion, the stream is diverted
and flows along the front of the range.

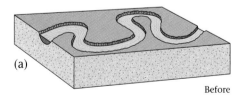

(a)

Before

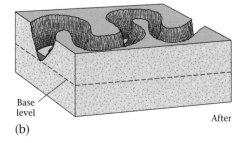

Base
level

After

(b)

FIGURE 17.31 (a) A stream forms meanders while it
flows across a plain. (b) Uplift of the land over which the
stream flows causes the meanders to cut down and carve
out canyons that meander like the stream.

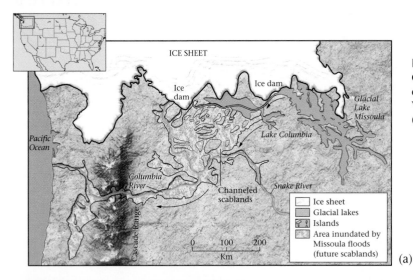

FIGURE 17.34 (a) The map illustrates Glacial Lake Missoula, blocked by an ice dam. (b) The channeled scablands, in Washington State, as viewed from the air. (c) A geologist's sketch of the photo.

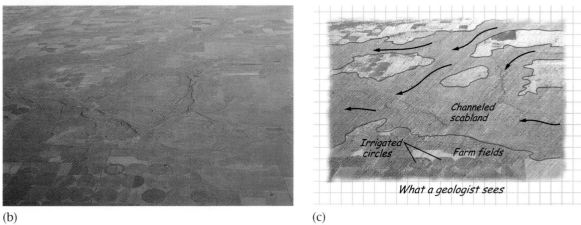

(b)

(c)

NOTES

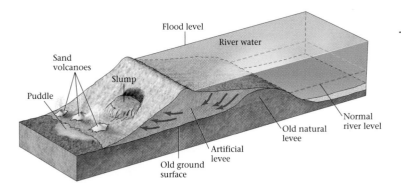

FIGURE 17.35 (a) When the water level on the river side of the levee is much higher than on the dry floodplain, pressure causes water to infiltrate the ground and flow through this artificial levee.

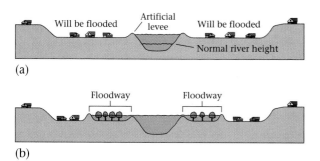

FIGURE 17.36 Concept of a floodway.

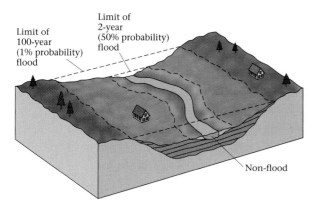

FIGURE 17.37 A 100-year flood covers a larger area than a 2-year flood.

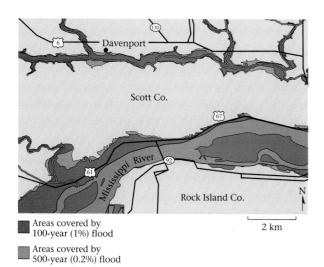

FIGURE 17.38 A flood hazard map for a region near Davenport, Iowa, as prepared by FEMA.

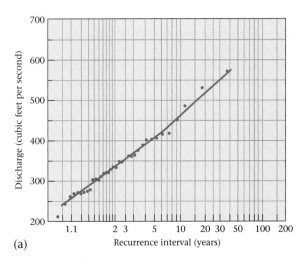

(a)

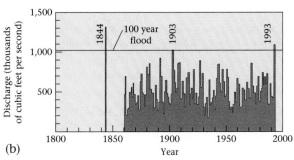

(b)

FIGURE 17.39 (a) A flood-frequency graph shows the relationship between the recurrence interval and discharge for an idealized river. (b) The peak discharge of the Mississippi River as measured at St. Louis, Missouri.

NOTES

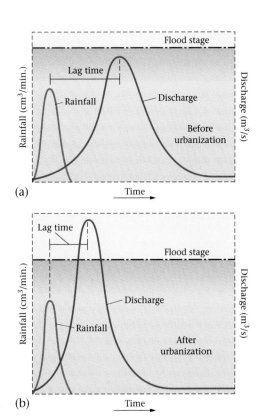

FIGURE 17.41 Hydrographs. (a) Before urbanization. (b) With urbanization.

CHAPTER 18 *Restless Realm: Oceans and Coasts*

FIGURE 18.2 (b) In this traditional seismic-reflection profile of the sea floor, the darker lines represent the boundaries between sedimentary layers in the oceanic crust.

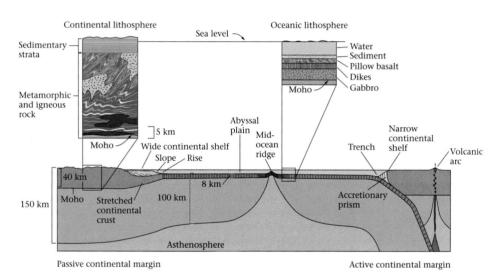

FIGURE 18.3 The bathymetric provinces of the sea floor.

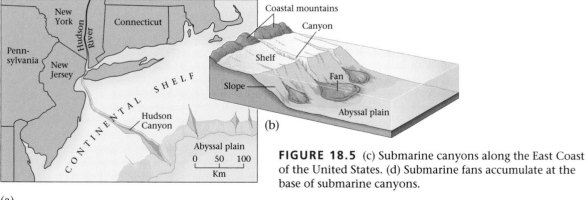

FIGURE 18.5 (c) Submarine canyons along the East Coast of the United States. (d) Submarine fans accumulate at the base of submarine canyons.

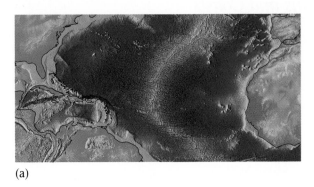

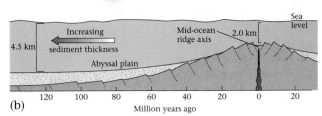

(a)

(b)

FIGURE 18.6 (a) A segment of a mid-ocean ridge, showing transform faults that link segments of the ridge. (b) The sea floor slopes away from a mid-ocean ridge and gradually flattens out to become an abyssal plain.

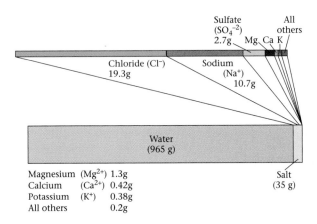

FIGURE 18.7 The composition of average seawater.

NOTES

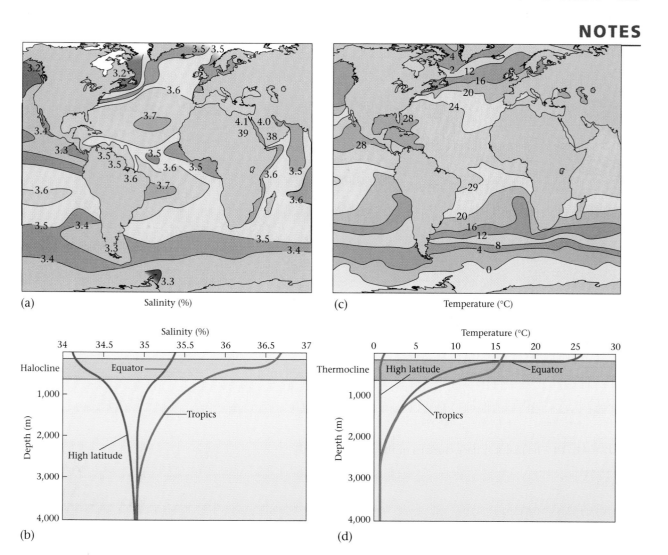

FIGURE 18.8 (a) The variations in salinity in the world ocean. (b) The variation of salinity with depth in the ocean. (c) The variation in temperature with latitude. Contours are given in degrees Celsius. (d) The variation in temperature with depth.

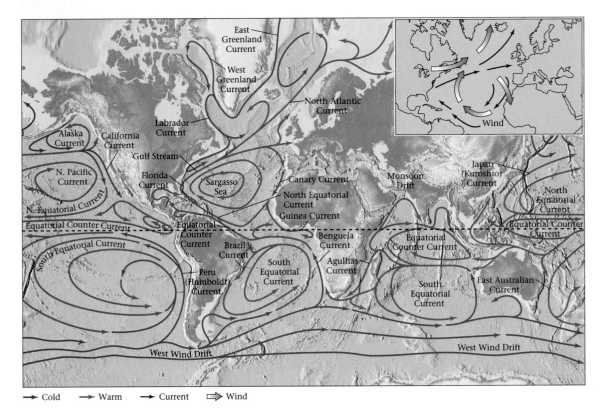

FIGURE 18.10 The major surface currents of the world's oceans.

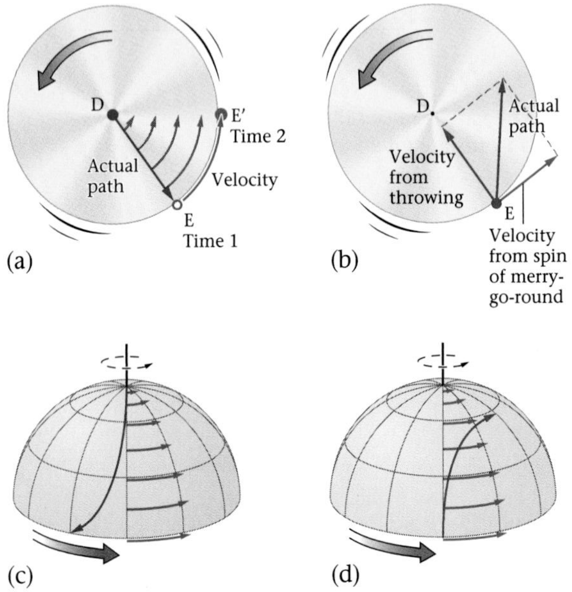

FIGURE 18.11 The Coriolis effect.

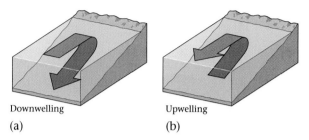

Downwelling   Upwelling

(a)   (b)

FIGURE 18.12 (a) Where surface water moves toward shore, it downwells to make room for more water. (b) Where surface water moves offshore, deep water upwells to replace the water that flowed away.

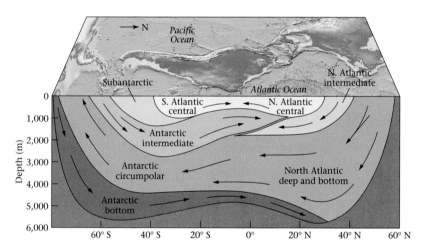

FIGURE 18.13 Because of variations in density, primarily caused by variations in temperature, the oceans are vertically stratified into moving water masses.

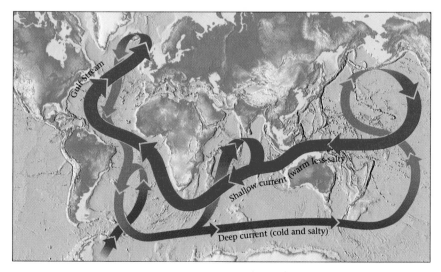

FIGURE 18.14 The exchange between upwelling deep water and downwelling surface water creates a global conveyor belt that circulates water throughout the entire ocean; this takes hundreds of years to millennia.

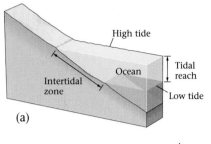

(a)

FIGURE 18.15 (a) The tidal reach is the difference between the high and low tide. (c) A larger tidal bulge appears on the side of the Earth closest to the Moon, and a smaller tidal bulge on the opposite side. (d) The tides as viewed looking down on the North Pole. (e) When the gravitational attraction of the Sun adds to that of the Moon, extra high tides, called spring tides, form.

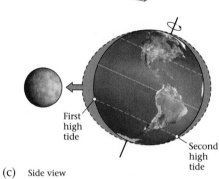

(c) Side view

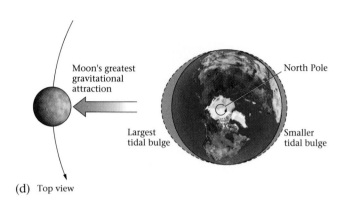

(d) Top view

NOTES

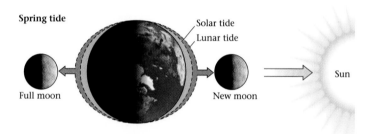

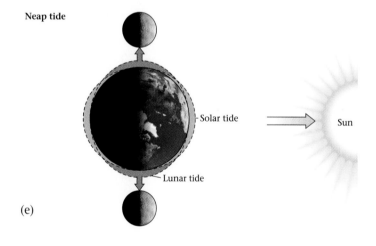

(e)

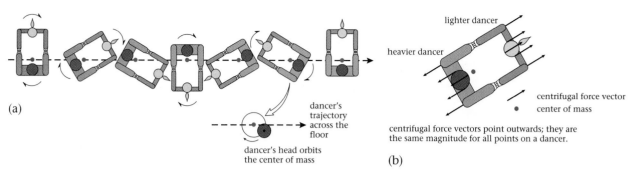

(a)

centrifugal force vectors point outwards; they are
the same magnitude for all points on a dancer.

(b)

FIGURE 18.16 (a) To picture the Earth-Moon system, imagine two dancers spinning around each
other as they move along a straight-line trajectory. (b) Each point on each dancer feels a centrifugal
force (represented by a vector) that points outward.

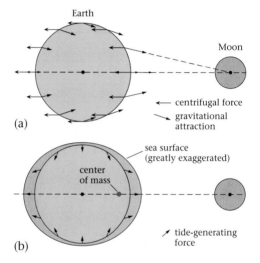

FIGURE 18.17 (a) Each point on the surface of
the Earth feels the same centrifugal force, but feels a
different gravitational attraction. (b) Tide-generating
force is the sum of the centrifugal force vector and
the gravitational force vector.

NOTES

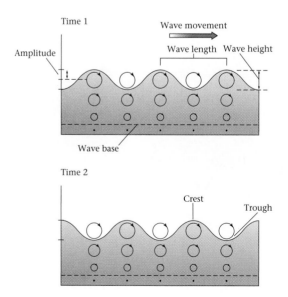

FIGURE 18.18 Within a deep-ocean wave, water
molecules follow a circular path.

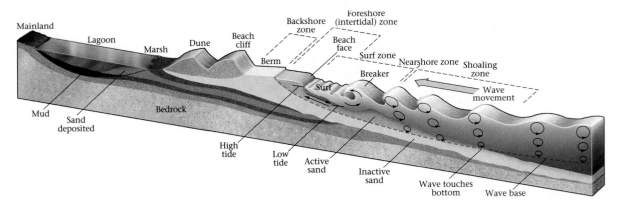

FIGURE 18.19 This profile shows the various landforms of a beach, as well as a cross section of a barrier island.

NOTES

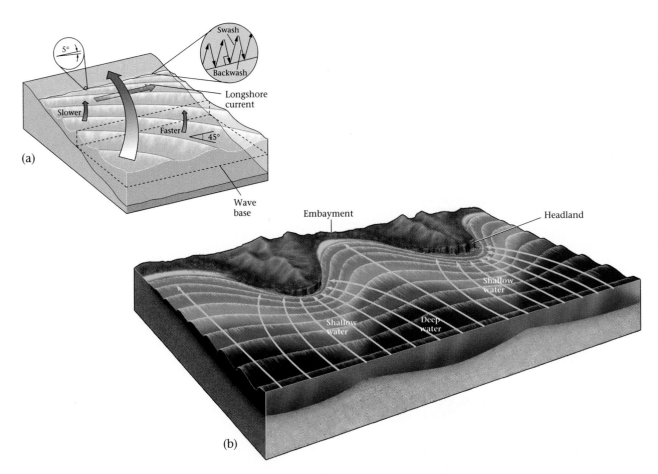

FIGURE 18.20 (a) Wave refraction occurs when waves approach the shore at an angle. (c) Like a lens, wave refraction focuses wave energy on a headland, so erosion occurs; and it disperses wave energy in embayments, so deposition occurs.

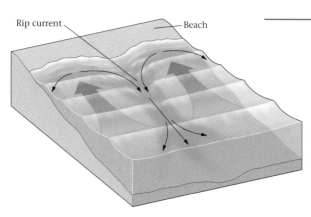

FIGURE 18.21 Waves bring water up on shore.

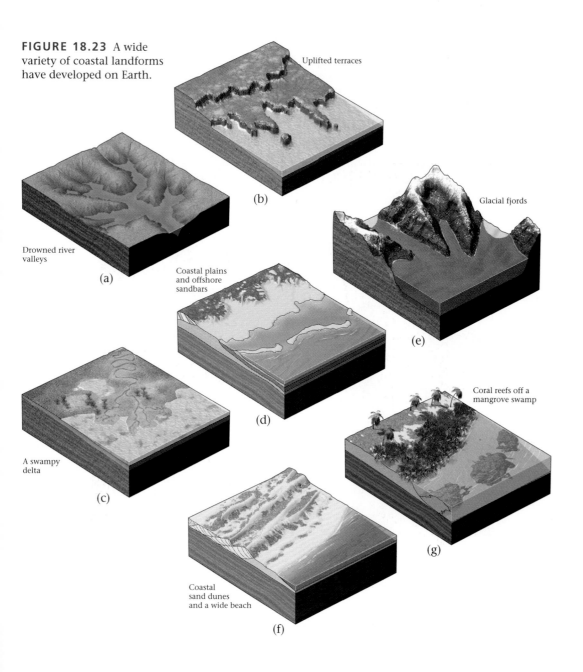

FIGURE 18.23 A wide variety of coastal landforms have developed on Earth.

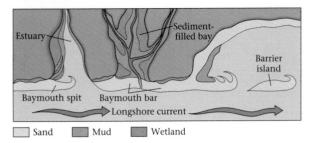

Estuary — Sediment-filled bay

Barrier island

Baymouth spit   Baymouth bar

Longshore current

☐ Sand   ▨ Mud   ▧ Wetland

FIGURE 18.25 Beach drift can generate sand spits and baymouth bars.

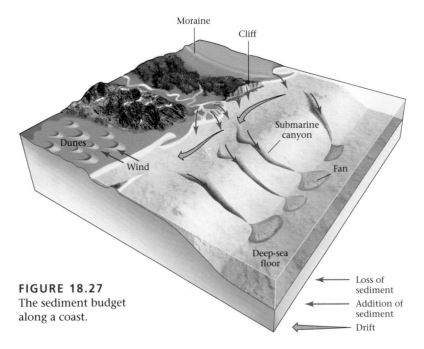

Moraine

Cliff

Dunes

Wind

Submarine canyon

Fan

Deep-sea floor

Loss of sediment

Addition of sediment

Drift

FIGURE 18.27
The sediment budget along a coast.

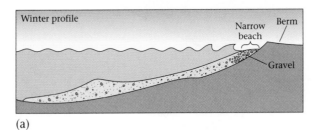

Winter profile

Narrow beach

Berm

Gravel

(a)

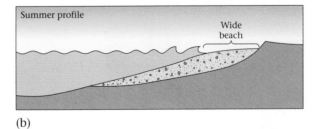

Summer profile

Wide beach

(b)

FIGURE 18.28 (a) In the winter, when waters are stormier, sand moves offshore, and the beach narrows and may become stonier. (b) During the summer, waves bring sand back to replenish the beach.

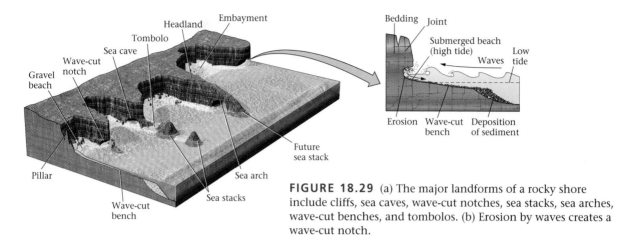

FIGURE 18.29 (a) The major landforms of a rocky shore include cliffs, sea caves, wave-cut notches, sea stacks, sea arches, wave-cut benches, and tombolos. (b) Erosion by waves creates a wave-cut notch.

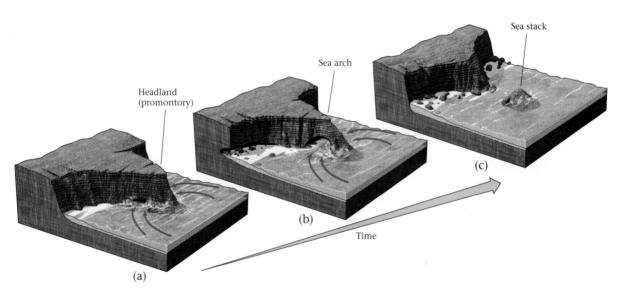

FIGURE 18.30 The erosion of a headland.

NOTES

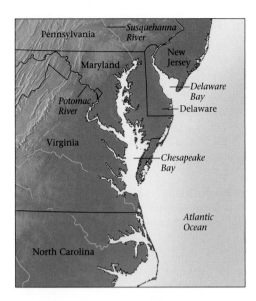

FIGURE 18.33 Chesapeake Bay, a large estuary along the East Coast of the United States, formed when sea level rose and flooded the Potomac and Susquehanna river valleys.

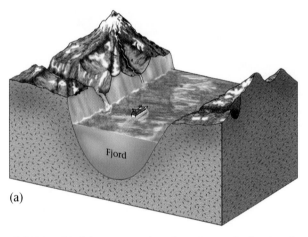

(a)

FIGURE 18.34 (a) The subsurface shape of a fjord, a drowned U-shaped glacial valley.

(c)

FIGURE 18.35 (c) The distribution of coral reefs on Earth today.

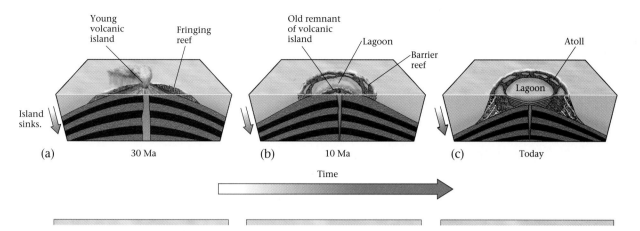

FIGURE 18.36 The progressive change from a fringing reef around a young volcanic island to a ring-shaped atoll.

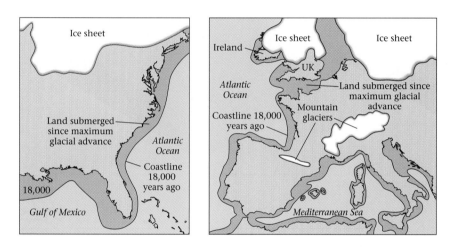

FIGURE 18.37 Exposed land of North America and Europe during the last ice age.

NOTES

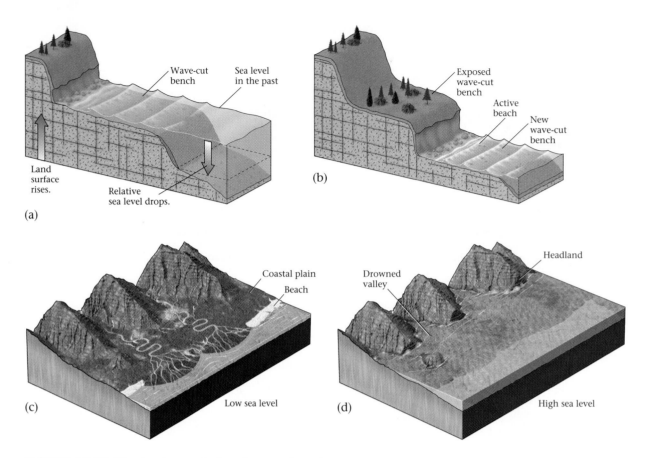

FIGURE 18.38 The development of a submergent coast.

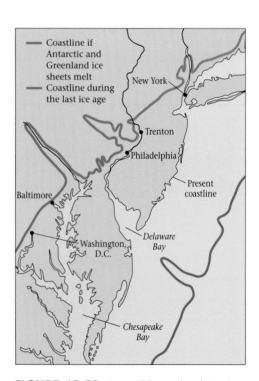

FIGURE 18.39 A possible sea-level rise in the future may flood major cities of the northeastern United States.

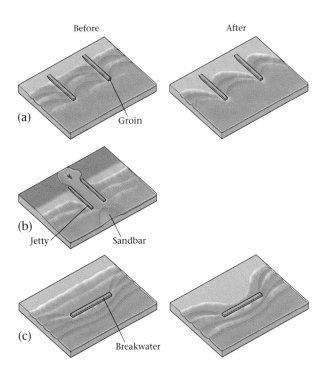

FIGURE 18.41 (a) The construction of groins creates a sawtooth beach. (b) Jetties extend a river farther into the sea, but may result in the deposition of a sandbar at the jetties' ends. (c) A breakwater causes the beach to build out in the lee.

NOTES

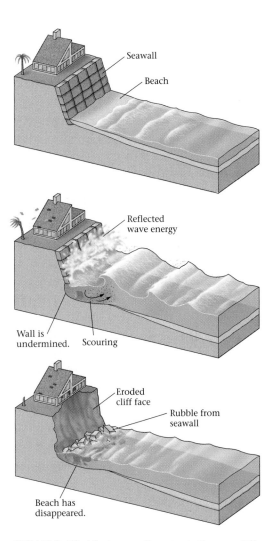

FIGURE 18.42 A seawall protects the sea cliff under most conditions, but during a severe storm the wave energy reflected by the seawall helps scour the beach.

CHAPTER 19 *A Hidden Reserve: Groundwater*

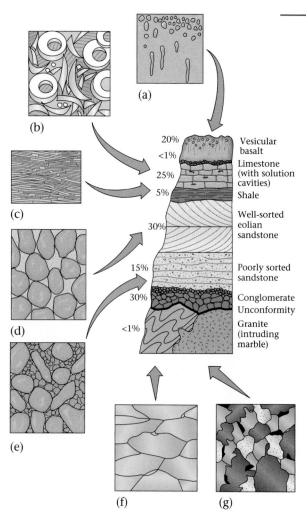

FIGURE 19.2 Various kinds of primary porosity in rock. Porosities are indicated as percentages.

FIGURE 19.3 Fractures in a rock provide secondary porosity.

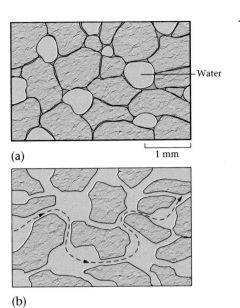

(a)

1 mm

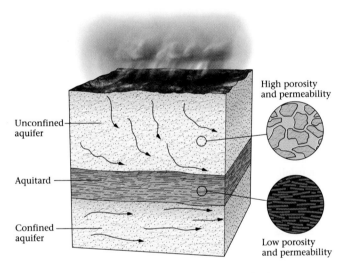

(b)

FIGURE 19.4 (a) Isolated, nonconnected pores in an impermeable material; (b) pores connected to one another by a network of conduits in permeable material.

High porosity and permeability

Unconfined aquifer

Aquitard

Confined aquifer

Low porosity and permeability

FIGURE 19.5 An aquifer is a high-porosity, high-permeability rock.

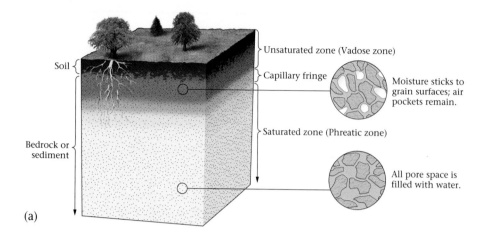

Soil

Bedrock or sediment

Unsaturated zone (Vadose zone)

Capillary fringe

Saturated zone (Phreatic zone)

Moisture sticks to grain surfaces; air pockets remain.

All pore space is filled with water.

(a)

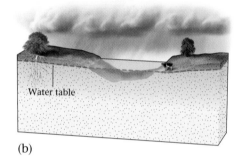

Water table

(b)

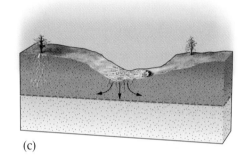

(c)

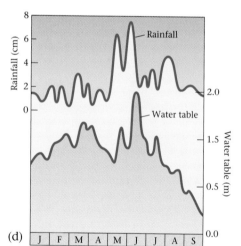

(d)

FIGURE 19.6 (a) The geometry of the water table, illustrating the saturated zone, the unsaturated zone, and the capillary fringe. (b) The surface of a permanent pond is the water table. (c) During the dry season, the water table can drop substantially, causing the pond to dry up. (d) The graph shows the relation between the height of the water table and rainfall between January and September, in a temperate region.

NOTES

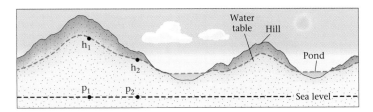

FIGURE 19.7 The shape of a water table beneath hilly topography.

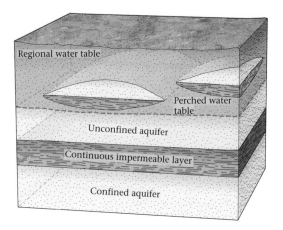

FIGURE 19.8 The configuration of a perched water table.

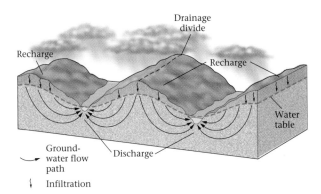

FIGURE 19.9 The flow lines from the recharge area to the discharge area curve through the substrate.

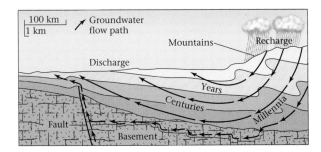

FIGURE 19.10 Cross section showing regional-scale groundwater flow in a sedimentary basin.

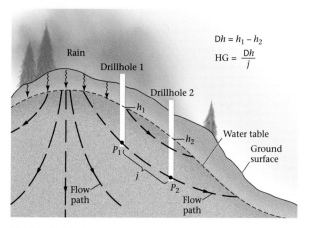

$$Dh = h_1 - h_2$$

$$HG = \frac{Dh}{j}$$

FIGURE 19.11 A hydraulic gradient (HG) is the change in hydraulic head per unit of distance between two points along the flow path.

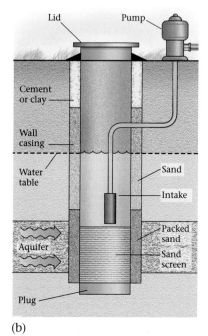

(b)

FIGURE 19.12 (b) Diagram showing how drillers set up a pump to extract water from an ordinary well.

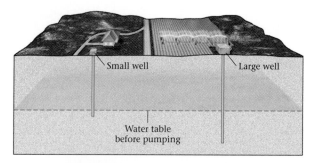

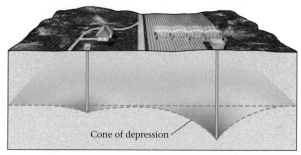

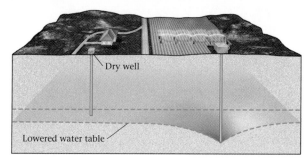

FIGURE 19.13 The base of an ordinary well penetrates below the water table.

NOTES

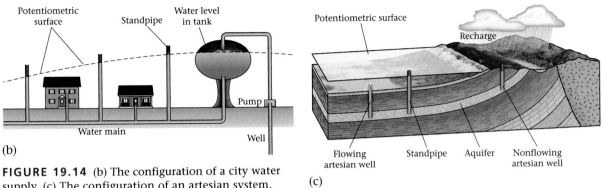

(b)

FIGURE 19.14 (b) The configuration of a city water supply. (c) The configuration of an artesian system.

(c)

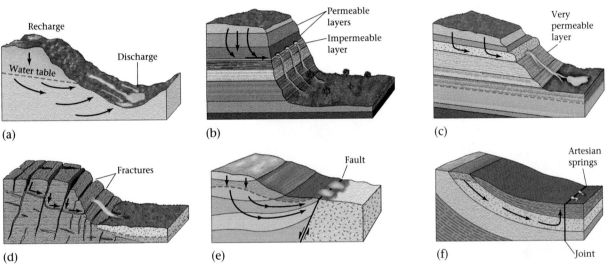

(a)

(b)

(c)

(d)

(e)

(f)

(g)

FIGURE 19.15 Springs form (a) where groundwater rises in a discharge area; (b) where groundwater has been forced to migrate along an impermeable barrier; (c) where a particular permeable layer transmits water to the surface of a hill; (d) where a network of interconnected fractures channels water to the hill face; (e) where groundwater collides with a steep impermeable barrier, and pressure pushes it up to the ground along the barrier. (f) An artesian spring forms where water from a confined aquifer migrates up a joint; (g) springs also form where a perched water table intersects the surface of a hill.

NOTES

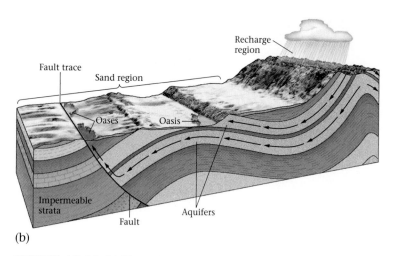

(b)

FIGURE 19.16 (b) This subsurface configuration of aquifers leads to the formation of an oasis, where groundwater reaches the surface.

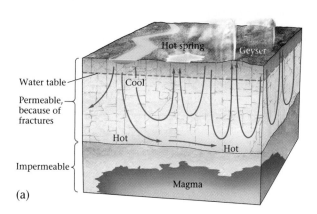

(a)

FIGURE 19.19 (a) How geysers and hot springs form .

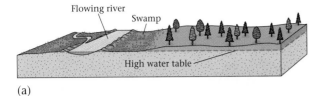

(a)

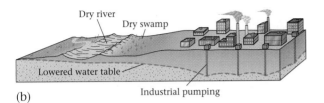

(b)

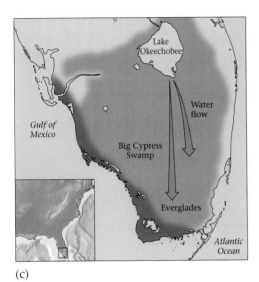

(c)

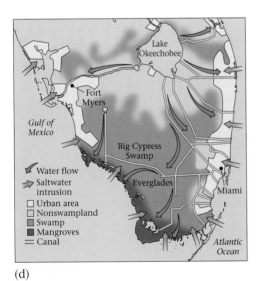

(d)

FIGURE 19.20 (a) Before a water table is lowered, a large swamp exists. (b) Pumping by a nearby city causes the water table to sink, so the swamp dries up. (c) The Everglades, in Florida, before the advent of urban growth and intensive agriculture. (d) Channelization and urbanization have removed water from the recharge area, disrupting the groundwater flow path in the Everglades.

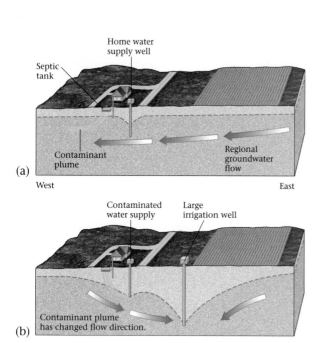

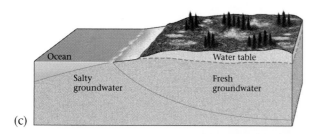

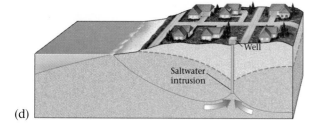

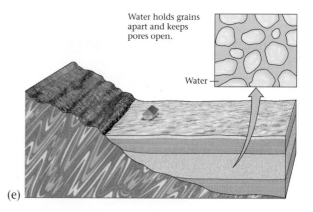

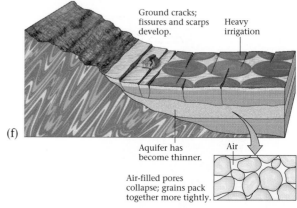

FIGURE 19.21 (a) Before pumping, effluent from a septic tank drifts west with the regional groundwater flow. (b) After pumping, it drifts east into the well, in response to the local slope of the water table. (c) Before pumping, fresh groundwater forms a large lens over salty groundwater. (d) Pumping too fast sucks saltwater from below into the well. (e, f) Pore space collapses when water is removed.

NOTES

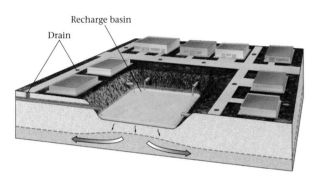

FIGURE 19.22 Sketch of an enhanced recharge catchment in a city.

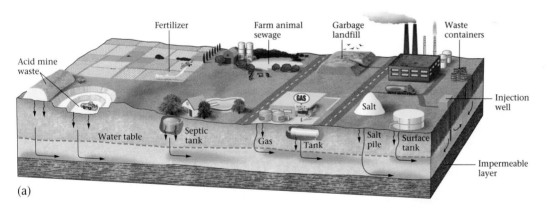

(a)

FIGURE 19.23 (a) The various sources of groundwater contamination. (b) A contaminant plume as seen in cross section. (c) A map of the contaminant plume emanating from a sewage-treatment basin of a military base on Cape Cod, Massachusetts.

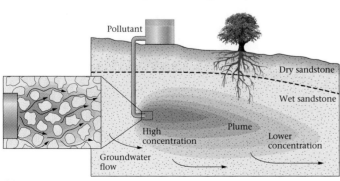

(b)

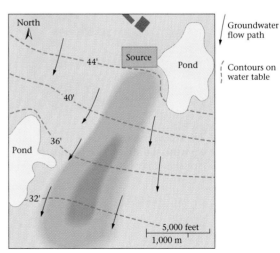

(c)

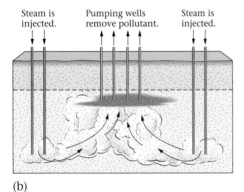

(a)

(b)

FIGURE 19.24 (a) Leaky drums of chemicals introduce pollutants into the groundwater. (b) Steam injected beneath the contamination drives the contaminated water upward in the aquifer, where pumping wells remove it.

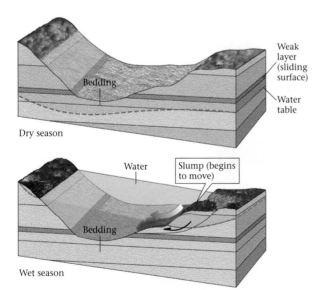

FIGURE 19.25 When the water table rises, material above a weak sliding surface begins to slump, and a landslide may result.

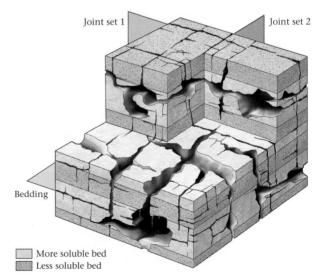

FIGURE 19.26 Joints act as conduits for water in cave networks. Thus, caves and passageways lie along joints.

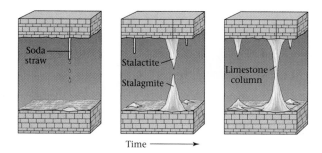

FIGURE 19.28 The evolution of a soda straw stalactite into a limestone column.

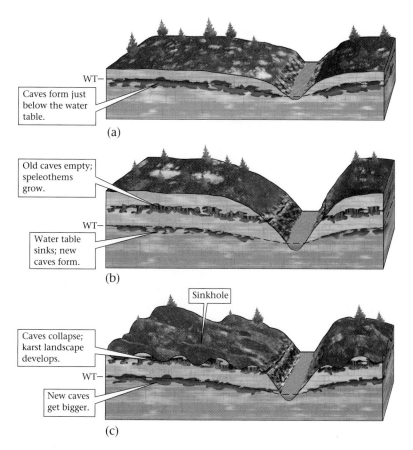

FIGURE 19.30 The formation of caves and a karst landscape.

CHAPTER 20 *An Envelope of Gas: Earth's Atmosphere and Climate*

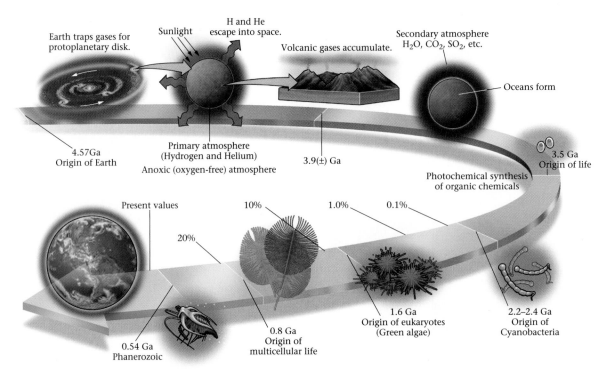

FIGURE 20.2 Stages in the evolution of the atmosphere.

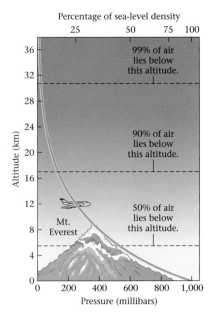

FIGURE 20.4 This graph shows air pressure versus elevation on the Earth.

NOTES

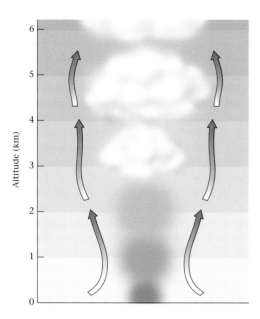

FIGURE 20.5 As air rises and enters regions of lower pressure, it expands and adiabatically cools.

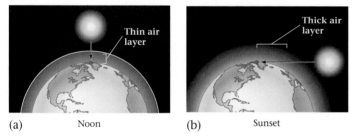

(a)   Noon   (b)   Sunset

FIGURE 20.6 (a) Scattering of light by gas molecules leads to the brilliant blue of a clear sky at noon. (b) So much scattering happens when sunlight hits the atmosphere at a low angle that all that's left is red light.

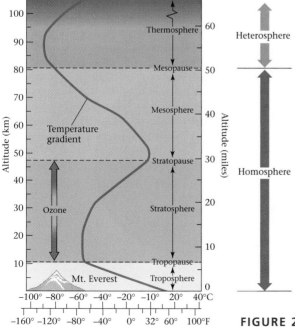

FIGURE 20.7 The principal layers of the atmosphere are separated from each other by pauses.

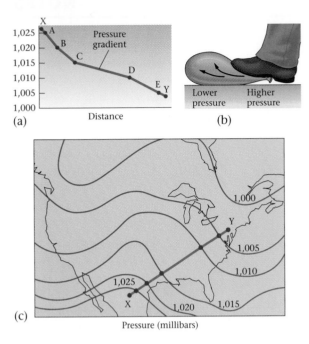

(a)

(b)

(c)

FIGURE 20.9 (a) This graph shows a profile from location X to location Y. (b) Air flows from a high-pressure region to a low-pressure region to cause wind. (c) Isobars on a map are lines of equal pressure.

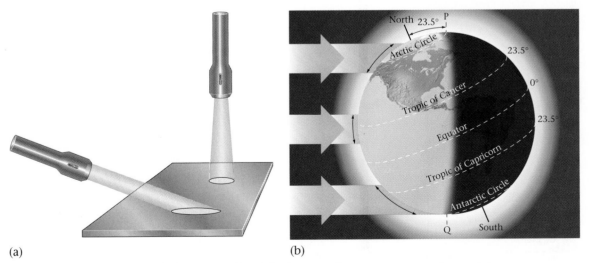

(a)

(b)

FIGURE 20.10 (a) A flashlight beam aimed straight down produces a narrower and more intense beam than a flashlight aimed obliquely. (b) Sunlight hitting the Earth near the equator provides more heat per unit area of surface than sunlight hitting the Earth at a polar latitude.

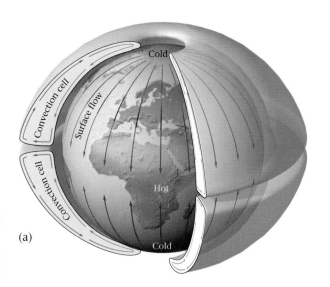

(a)

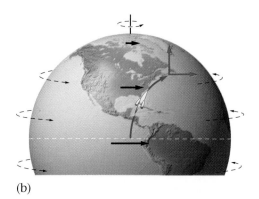

(b)

FIGURE 20.11 (a) If the Earth did not rotate, two simple convection cells would be established in the atmosphere, one stretching from the equator to the North Pole and one stretching from the equator to the South Pole. (b) Because of the Coriolis effect, a rocket sent north from the equator curves to the right (east). (c) Because of the Coriolis effect, atmospheric circulation breaks into three convection cells within each hemisphere, named, from equator to pole, Hadley, Ferrel, and polar.

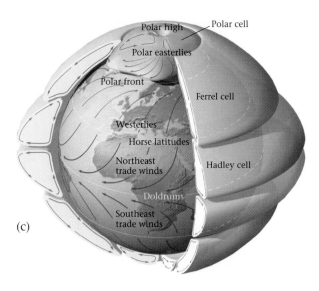

(c)

NOTES

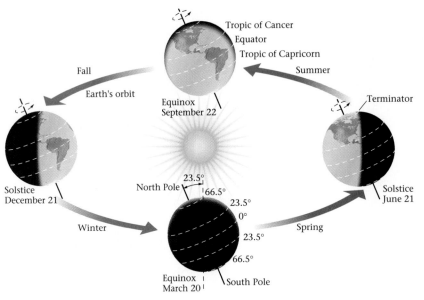

FIGURE 20.12 Because of the tilt of the Earth's axis, we have seasons.

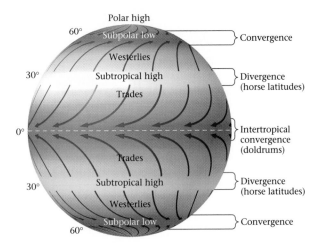

FIGURE 20.13 If the Earth had a uniform surface, distinct high- and low-pressure zones would form on its surface.

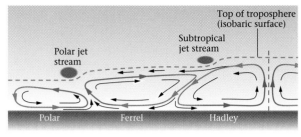

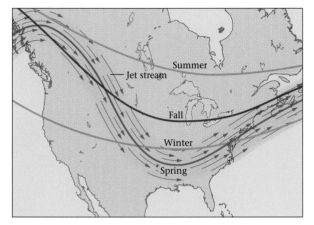

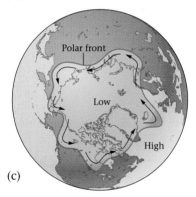

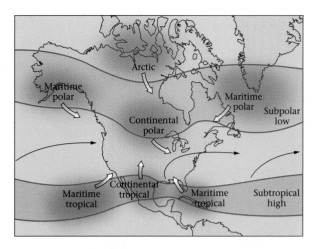

FIGURE 20.15 Air masses that form in different places have been assigned different names.

NOTES

FIGURE 20.14 (a) In an equator-to-pole cross section of the atmosphere, we see that the troposphere varies in thickness. (b) The polar jet stream meanders with time, and changes its overall position with the seasons; it tends to be farther south in the winter than in the summer. (c) Looking down on the Earth from the North Pole, we see the irregular shape of the polar front, along which the polar jet stream flows.

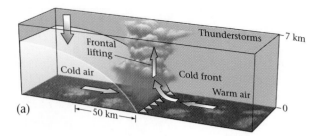

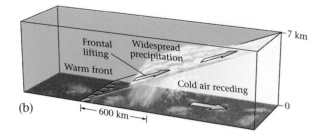

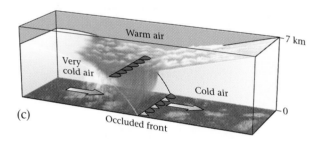

FIGURE 20.16 (a) A cold front develops where a cold air mass moves under a warm air mass. (b) A warm front develops where a warm air mass moves into a cold one. (c) An occluded front develops where a fast-moving cold front overtakes a warm front and lifts the base of the warm front off the ground.

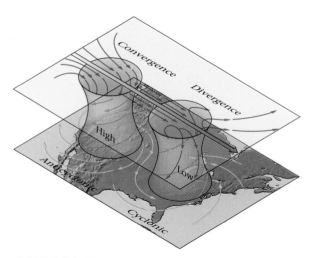

FIGURE 20.17 Air spirals downward and clockwise (creating an anticyclone) at a high-pressure mass and upward and counterclockwise (creating a cyclone) at a low-pressure mass.

(a)

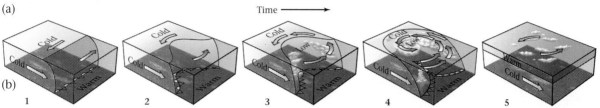

Time ⟶

(b)   1   2   3   4   5

FIGURE 20.18 (a) A fully developed mid-latitude (wave) cyclone in the middle of its west-to-east trek across North America. (b) The evolution of a mid-latitude cyclone.

NOTES

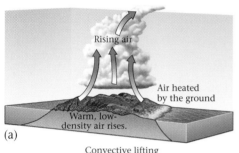

(a)   Convective lifting

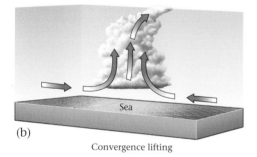

(b)   Convergence lifting

(c)   Orographic lifting

FIGURE 20.20 (a) Convective lifting occurs where particularly warm air starts to rise. (b) Convergence lifting takes place where winds merge—the air has nowhere to go but up. (c) Orographic lifting occurs where winds off the sea run into a mountain range and are forced up.

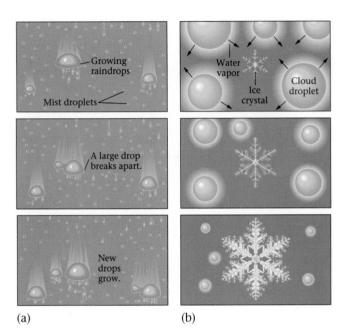

(a)   (b)

FIGURE 20.21 (a) Some rain forms when droplets collide and coalesce. (b) During the Bergeron process, water drops evaporate, releasing vapor that attaches to growing snowflakes, which then fall.

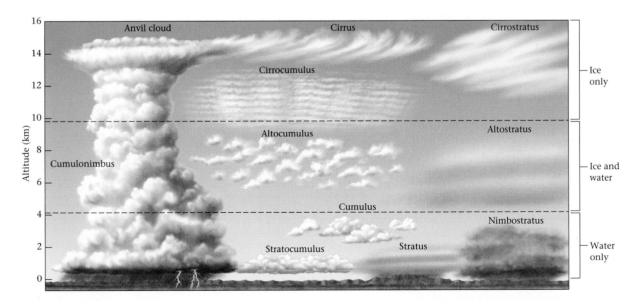

FIGURE 20.22 The type of cloud that forms in the sky depends on the stability of the air, the elevation at which moisture condenses, and the wind speed.

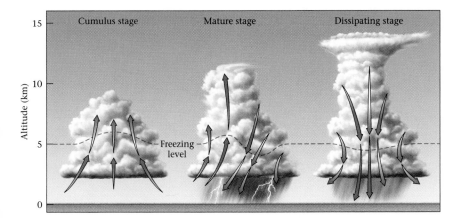

FIGURE 20.24 A thunderstorm evolves in three stages.

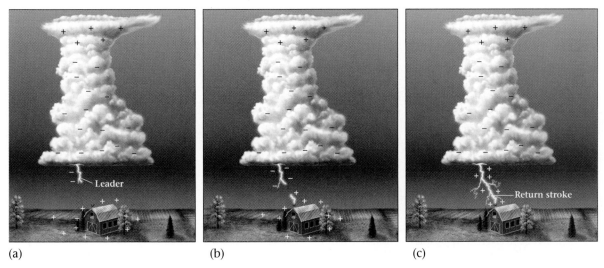

(a)   (b)   (c)

FIGURE 20.25 (a) Lightning flashes when a charge separation develops in a cloud, with a negative charge at the base and a positive charge at the top. (b) As the leader grows downward, positive charges begin to flow upward from an object on the ground. (c) When the connection is complete, the return stroke carries positive charges rapidly from the ground to the cloud, creating the main part of the flash.

NOTES

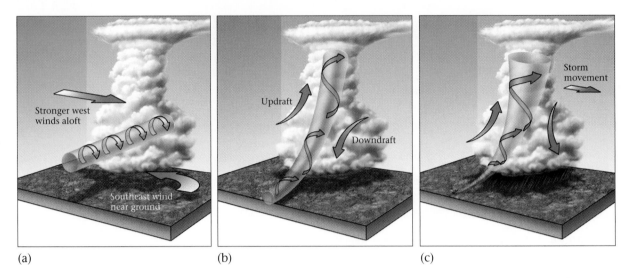

(a) (b) (c)

FIGURE 20.28 (a) Tornadoes initiate at intense fronts where high-altitude westerlies flow over low-altitude southeasterlies. (b) Updrafts and downdrafts in the cloud eventually tilt the cylinder, creating a tornado. (c) When the tornado touches down, destruction follows.

NOTES

Number of tornadoes per year (per 26,000 sq. km, for a 27-year period)

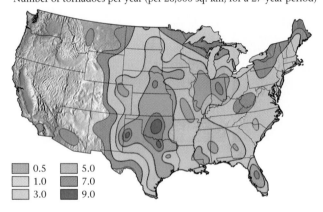

0.5 5.0
1.0 7.0
3.0 9.0

FIGURE 20.29 North American tornadoes are most common in "tornado alley," a band extending from Texas to Indiana.

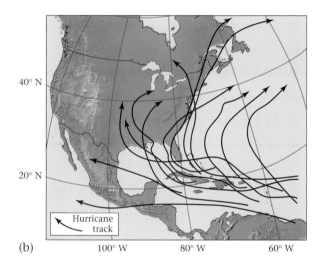

(b)

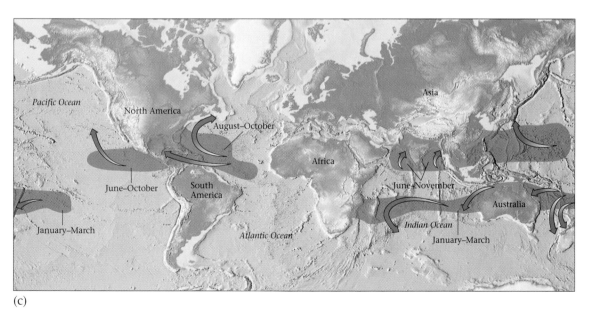

(c)

FIGURE 20.30 (b) The tracks of several important Atlantic hurricanes show how most begin at latitudes of 15°–20° off the western coast of Africa, then drift westward and northward. (c) Some Atlantic hurricanes make it across Central America and lash the eastern Pacific.

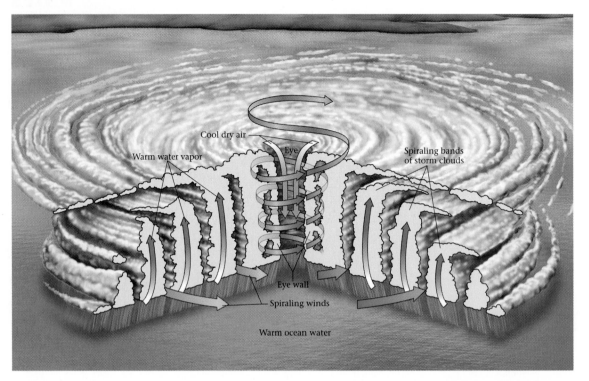

FIGURE 20.31 This cutaway diagram of a hurricane shows the spirals of clouds, the eye, and the eye wall.

NOTES

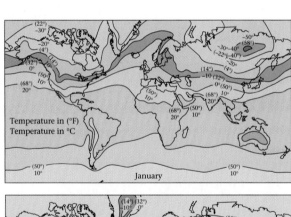

FIGURE 20.33 Isotherms of January roughly parallel lines of latitude. The pattern of isotherms is different in July.

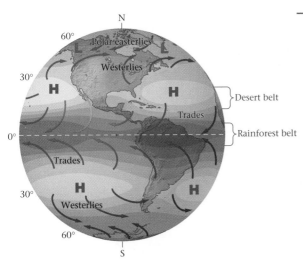

FIGURE 20.34 Because of land masses, atmospheric pressure belts vary in width to create lens-shaped, semipermanent high- and low-pressure cells.

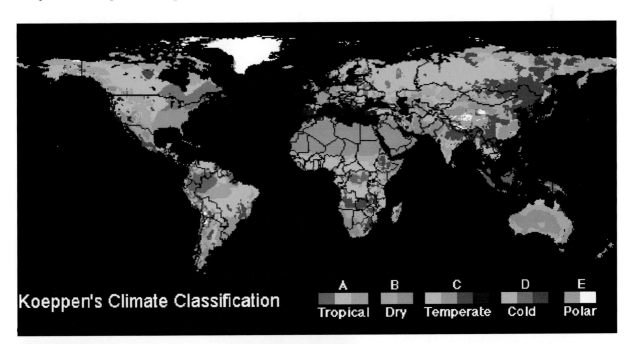

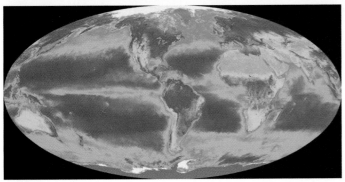

(b)

FIGURE 20.35 (a) The basic climate belts on Earth as originally defined by W. Koeppen (1846–1940). (b) Satellite image showing the global biosphere.

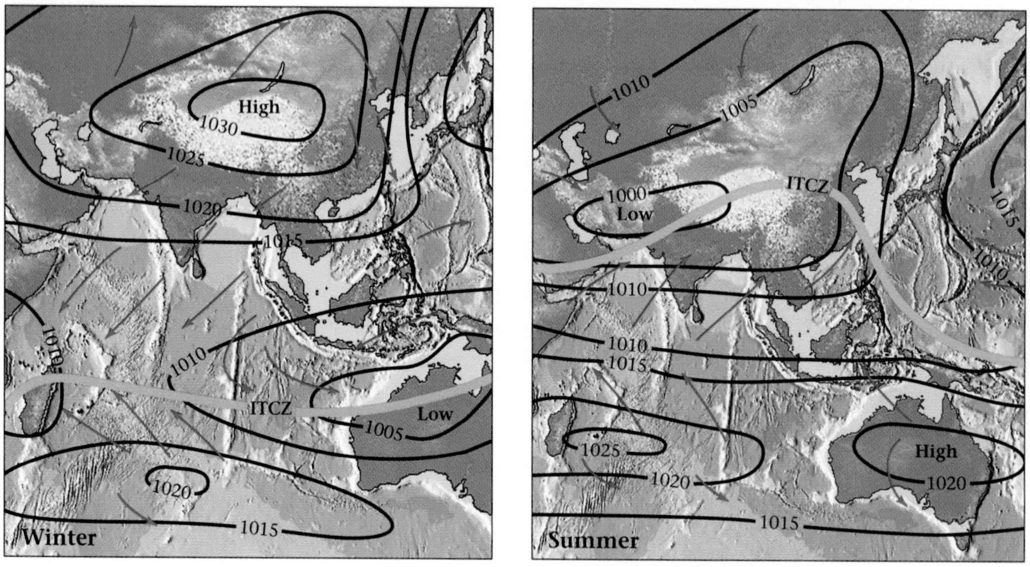

(a)

(b)

FIGURE 20.36 In the monsoonal climate of Asia, each year can be divided into a dry season and a wet season.

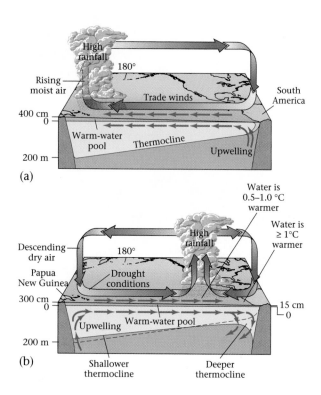

(a)

(b)

FIGURE 20.37 El Niño exists because of a change in winds and currents in the central Pacific.

NOTES

CHAPTER 21 *Dry Regions: The Geology of Deserts*

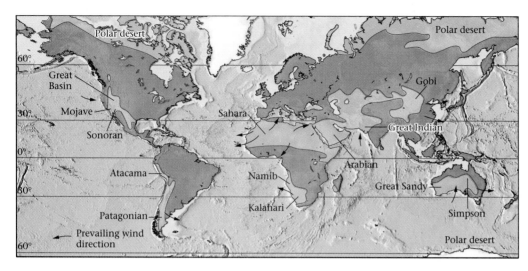

FIGURE 21.3 The global distribution of deserts.

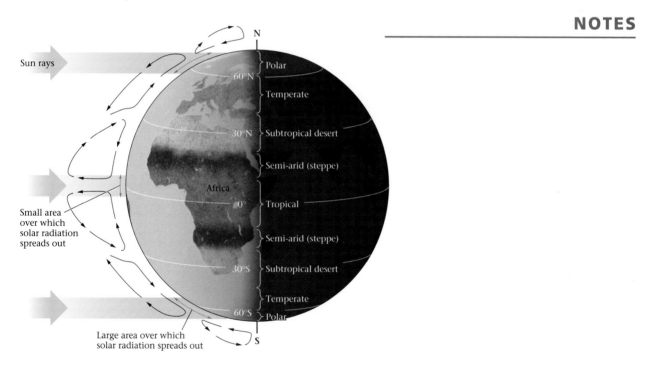

FIGURE 21.4 Rising air at the equator loses its moisture by raining over rainforests.

FIGURE 21.5 Moist air, when forced to rise by mountains, cools.

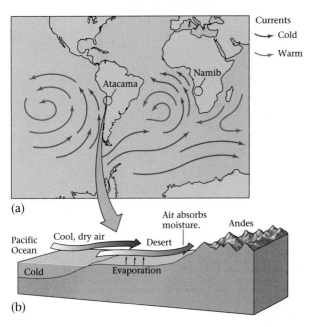

(a)

(b)

FIGURE 21.6 (a) Currents bringing cold water up from the Antarctic cool the air along the southwestern coasts of South America and Africa. (b) The cool, dry air absorbs moisture from the adjacent coastal land, keeping it dry, so coastal deserts form.

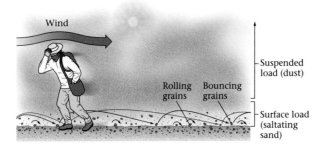

FIGURE 21.10 During saltation, sand grains roll and bounce along the ground surface.

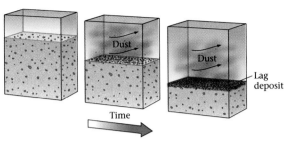

FIGURE 21.11 The progressive development of a lag deposit.

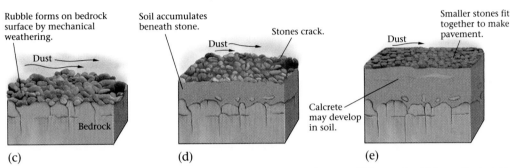

FIGURE 21.12 (c) Desert pavement forms in stages. First, loose pebbles and cobbles collect at the surface. (d) Dust settles among stones and builds up a layer of soil beneath the stone layer. (e) A durable, mosaic-like pavement has formed.

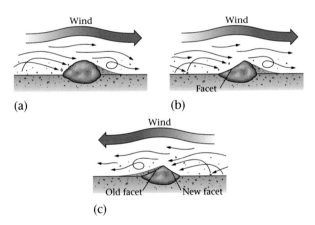

FIGURE 21.13 The progressive development of a ventifact.

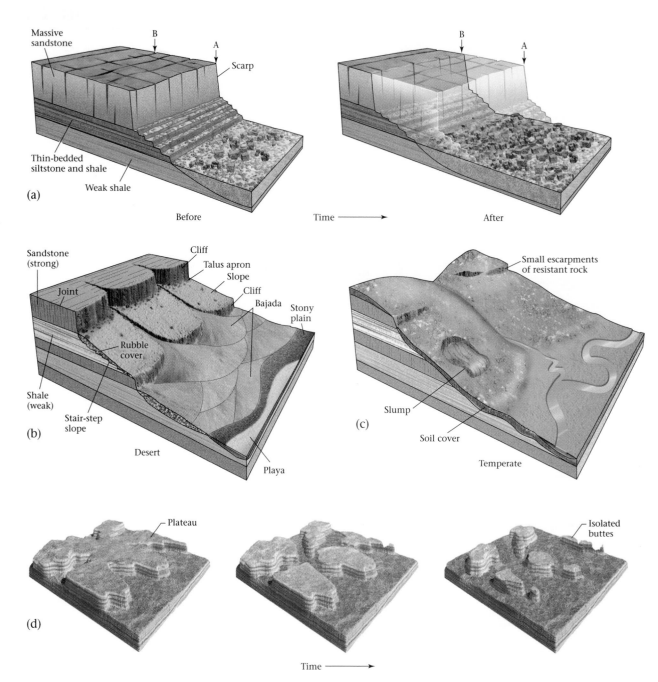

FIGURE 21.20 (a) The process of cliff retreat in a desert. (b) Stair-step cliffs appear where beds of strong rock are interlayered with thin beds of rock. (c) If the sequence of rocks shown in (b) were to occur in a temperate or humid climate, the slope would instead be smooth and underlain by thick soil. (d) Because of cliff retreat, a once-continuous layer of rock evolves into a series of isolated remnants.

NOTES

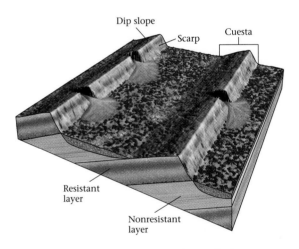

FIGURE 21.21 Asymmetric ridges called cuestas appear where the strata in a region are not horizontal.

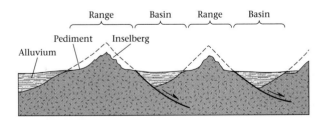

FIGURE 21.22 Inselbergs are small islands of rock surrounded by pediments.

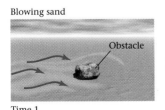

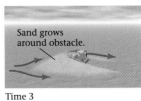
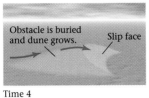

FIGURE 21.23 Progressive stages in the growth of a small sand dune.

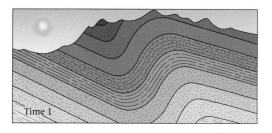

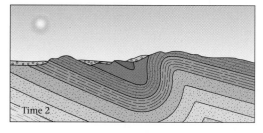

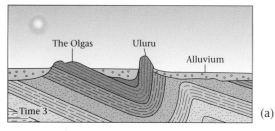

(a)

FIGURE 21.24 (a) Originally Uluru (Ayers Rock), in Australia, was part of a large syncline beneath a mountain.

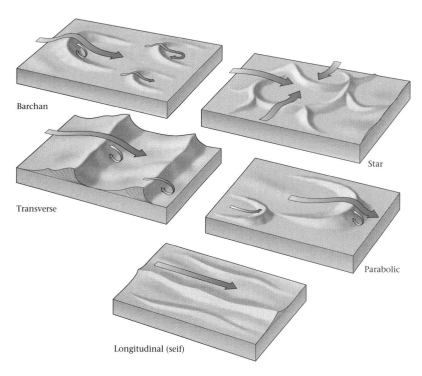

FIGURE 21.25 The various kinds of sand dunes.

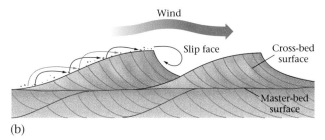

(b)

FIGURE 21.26 (b) Cross bedding inside a dune.

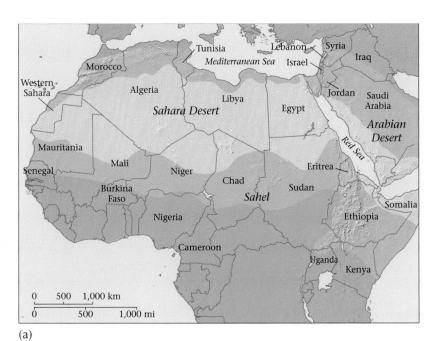

(a)

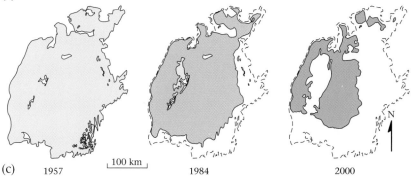

(c)   1957   1984   2000

FIGURE 21.29 (a) The Sahel is the semi-arid land along the southern edge of the Sahara Desert. (c) The Aral Sea in central Asia has started to dry up because of diversion of rivers that once flowed into it.

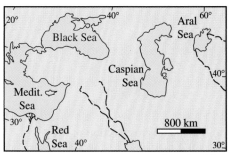

Location Map

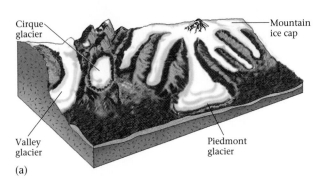

FIGURE 22.4 (a) The various kinds of mountain glaciers.

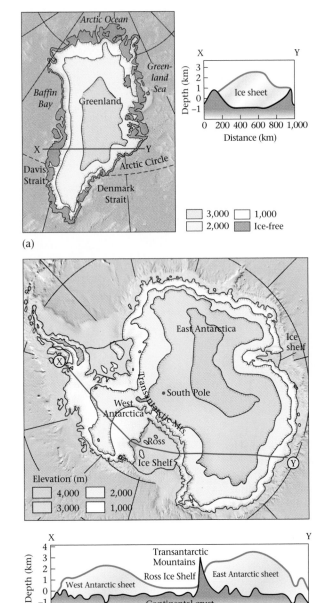

FIGURE 22.5 (a) A map and cross section of the Greenland ice sheet. (b) A map and cross section of the Antarctic ice sheet.

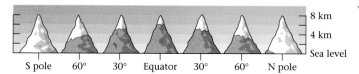

FIGURE 22.6 The snow line depends on latitude.

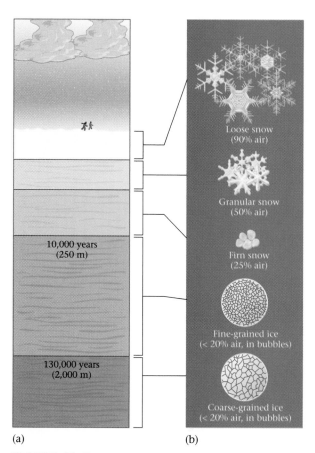

FIGURE 22.7 (a) Snow compacts and melts to form firn, which recrystallizes to make ice. (b) The size of ice crystals increases with depth in a glacier, where some crystals grow at the expense of others.

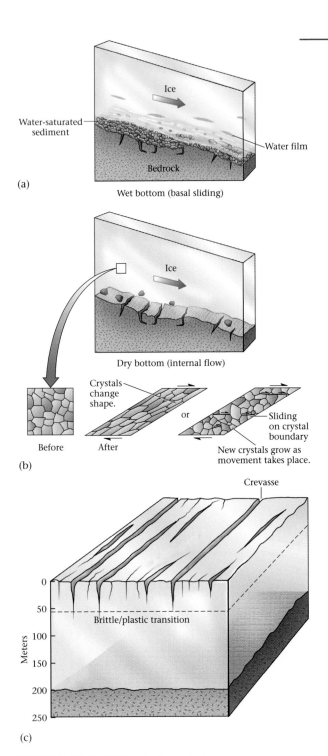

FIGURE 22.8 (a) Wet-bottom glaciers move by means of basal sliding. (b) Dry-bottom glaciers flow in the solid state (internally). (c) Crevasses form in the brittle ice above the brittle-plastic transition in glaciers.

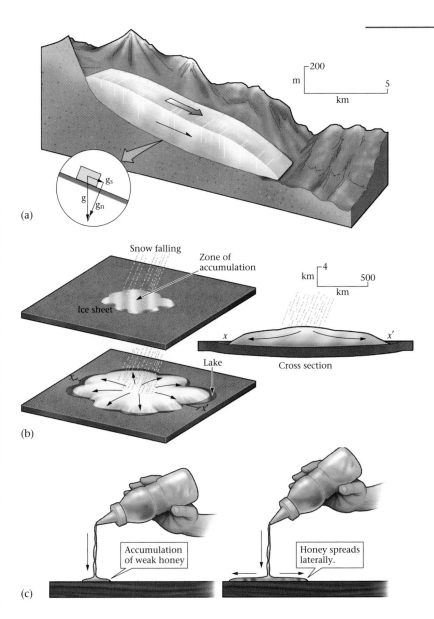

FIGURE 22.10 (a) The downslope motion of glaciers is driven by gravity, one component of which (g_s) is parallel to the slope. (b, c) The gravitational spreading of an ice sheet is similar to the spreading of honey.

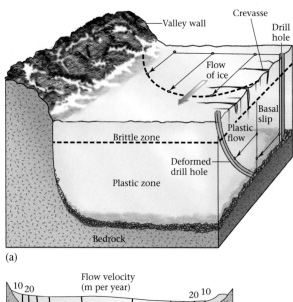

(a)

(b)

FIGURE 22.11 (a) Different parts of a glacier may flow at different velocities. (b) This cross section of a glacier shows measured velocities of flow.

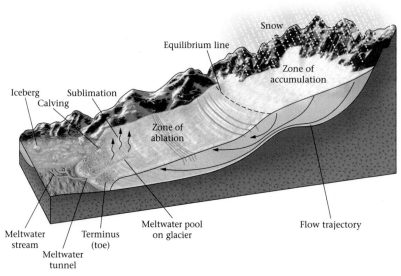

FIGURE 22.12 The zone of ablation, zone of accumulation, and equilibrium line.

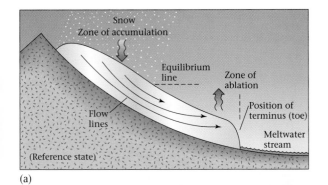

(a)

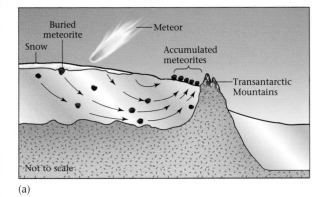

(b)

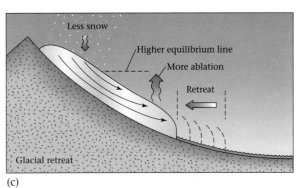

(c)

FIGURE 22.13 (a) The position of a glacier's toe represents the balance between the amount of ice that forms beneath the zone of accumulation and the amount of ice lost in the zone of ablation. (b) Glacial advance and (c) glacial retreat.

NOTES

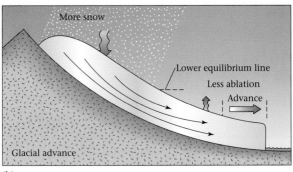

(a)

FIGURE 22.14 (a) Beneath the zone of ablation, ice crystals move up to the surface.

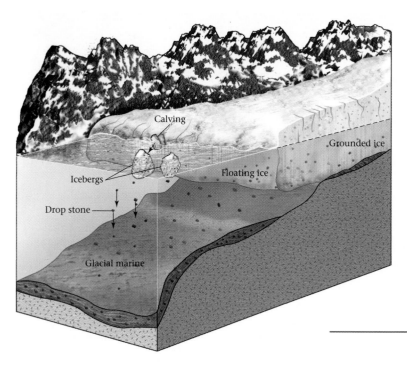

FIGURE 22.16 Where a glacier reaches the sea, the ice stays grounded in shallow water and floats in deep water.

NOTES

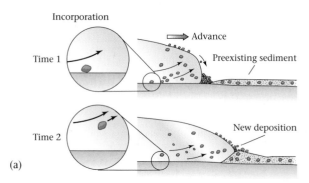

(a)

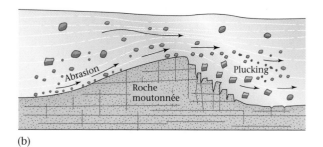

(b)

FIGURE 22.21 The various kinds of glacial erosion. (a) Plowing and incorporation; (b) plucking and abrasion.

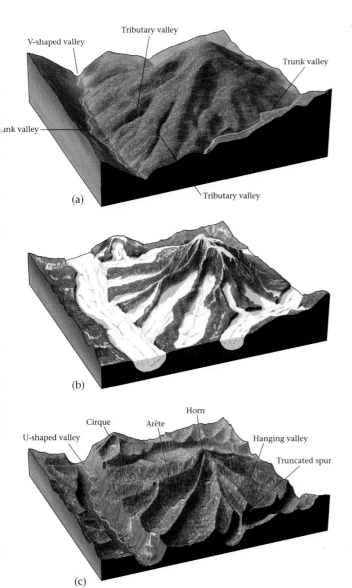

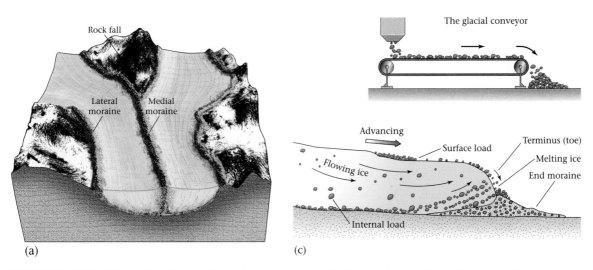

FIGURE 22.22 (a) A mountain landscape before glaciation. (b) During glaciation, the valleys fill with ice. (c) After glaciation, the region contains U-shaped valleys, hanging valleys, truncated spurs, and horns.

FIGURE 22.25 (a) Lateral and medial moraines on a glacier. (c) The surface load plus the internal load accumulates at the toe of the glacier to constitute an end moraine.

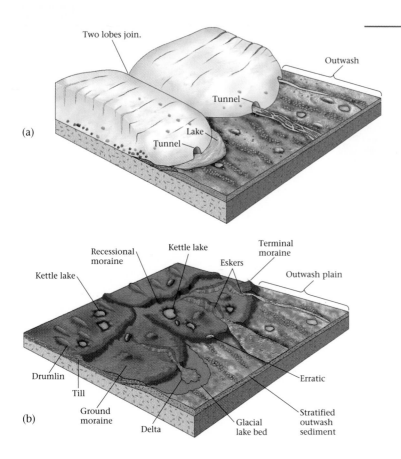

FIGURE 22.27 (a) The depositional landforms resulting from glaciation. (b) The setting in which various types of moraines form.

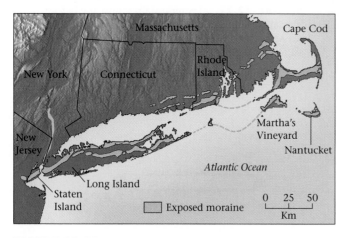

FIGURE 22.28 The moraines that constitute Long Island and Cape Cod.

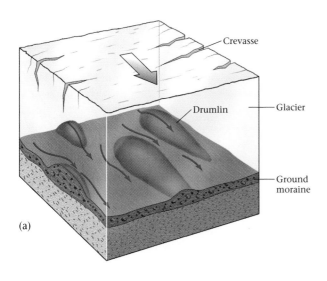

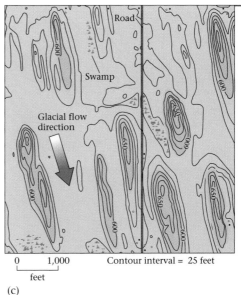

FIGURE 22.29 (a) The formation of a drumlin beneath a glacier. (c) Topographic map emphasizing the shape of drumlins in New York State.

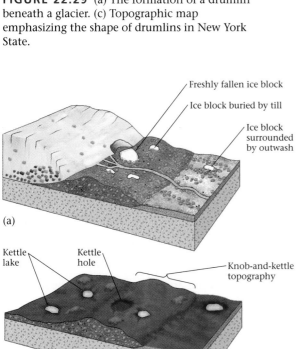

FIGURE 22.30 (a) When this ice block melts away, a kettle hole will form. (b) Knob-and-kettle topography, the hummocky surface of a moraine.

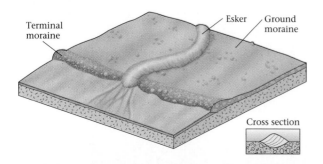

FIGURE 22.31 Eskers are snake-like ridges of sand and gravel that form when sediment fills meltwater tunnels at the base of a glacier.

NOTES

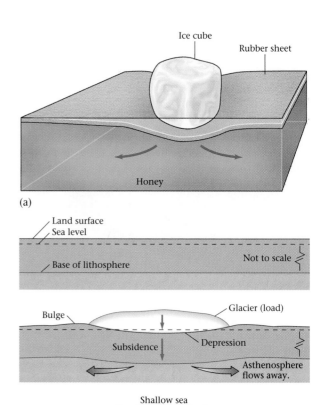

(a)

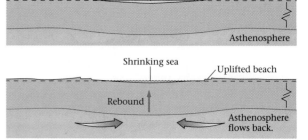

(b)

FIGURE 22.32 (a) An ice cube placed on the surface of a rubber sheet floating on honey illustrates the concept of glacial loading. (b) Cross sections illustrating the concept of glacial rebound.

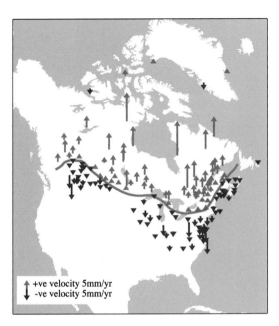

FIGURE 22.33 Rates of isostatic movement as indicated by satellite data.

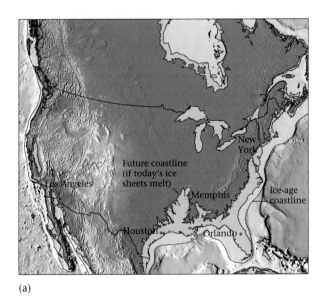

(a)

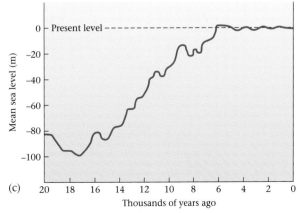

(b)

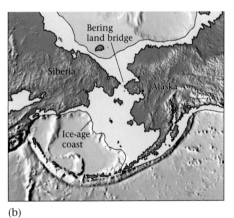

(c)

FIGURE 22.34 (a) This map shows the coastline of North America during the last ice age, and the coastline should present-day ice sheets melt. (b) A land bridge existed across the Bering Strait between Asia and North America during the last ice age. (c) The graph shows the change in sea level during the past 20,000 years.

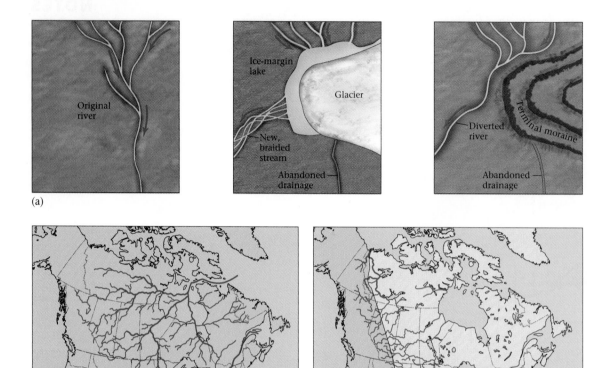

(a)

(b)

FIGURE 22.35 (a) When a glacier advances on the course of a river, the glacier blocks the drainage and causes a new stream to form. (b) In North America, the major river systems that flowed northward before the last ice age have been destroyed.

NOTES

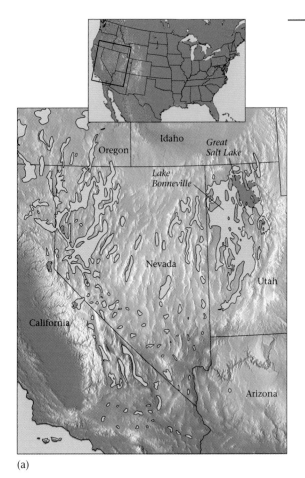

(a)

FIGURE 22.36 (a) The distribution of pluvial lakes in the Basin and Range Province during the last ice age. (b) The shoreline of Lake Bonneville along a mountain near Salt Lake City. The Great Salt Lake is a small remnant of this once-huge pluvial lake.

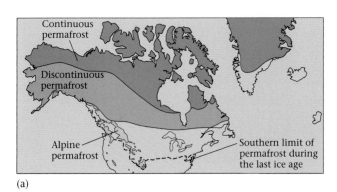

(a)

FIGURE 22.37 (a) The present-day distribution of periglacial environments in the Northern Hemisphere.

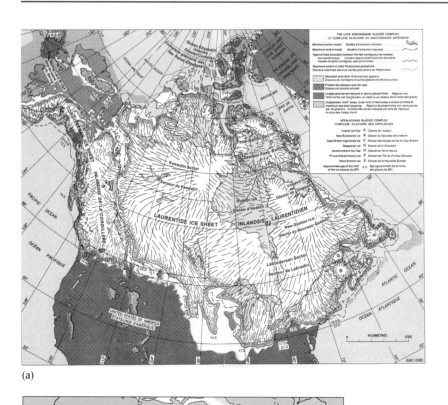

(a)

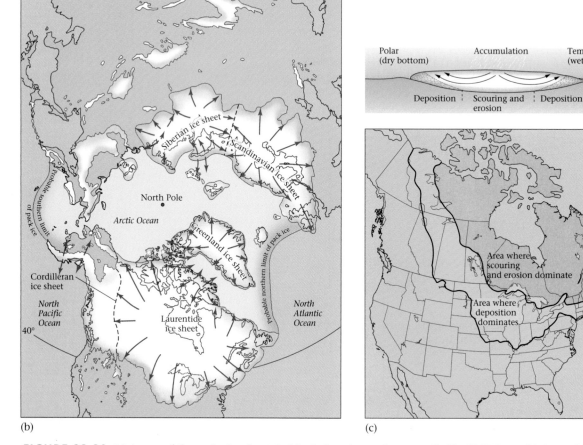

(b)

(c)

FIGURE 22.39 (a) A map of the major ice sheets in North America, as first compiled by V. K. Prest. (b) A simplified map of the distribution of major ice sheets during the Pleistocene Epoch. (c) Top: A continental glacier scours and erodes the land surface beneath its center, while at the margins it deposits sediment. Map: The Laurentide ice sheet scoured and eroded the land in northern and eastern Canada, and deposited sediment in western Canada and the midwestern United States.

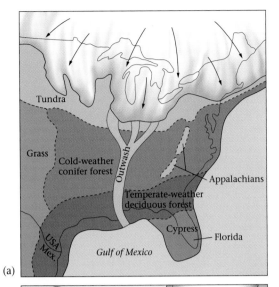

(a)

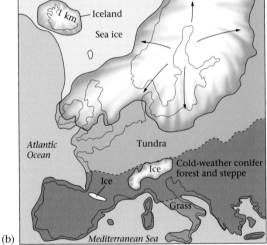

(b)

FIGURE 22.40 (a) Pleistocene climatic belts in North America; (b) Pleistocene climatic belts in Europe.

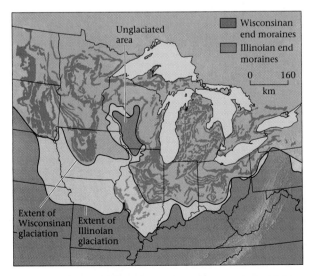

FIGURE 22.42 Pleistocene deposits in the United States.

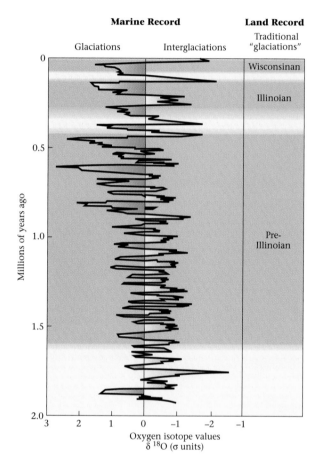

FIGURE 22.43 This time column shows the variations in oxygen-isotope ratios from marine sediment that define twenty to thirty glaciations and interglacials during the Pleistocene Epoch.

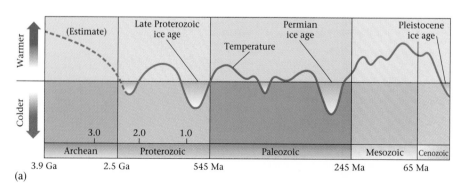

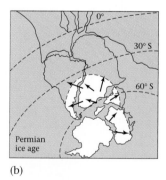

FIGURE 22.44 (a) This time column shows pre-Pleistocene glaciations during Earth history. (b) The distribution of Permian glacial features on a reconstruction of Gondwana, the southern part of the supercontinent that existed at the time.

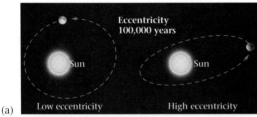

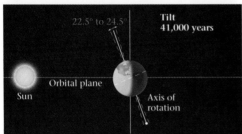

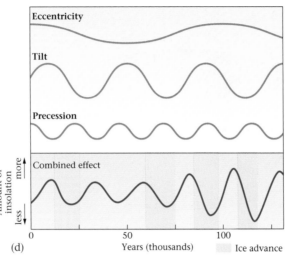

(a)

(b)

(c)

(d)

FIGURE 22.45 The Milankovitch cycles affect the amount of insolation (exposure to the Sun's rays) at high latitudes.

NOTES

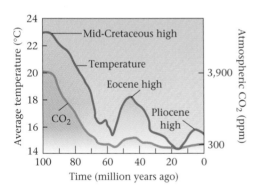

FIGURE 22.46 The graph shows the gradual cooling of Earth's atmosphere since the Cretaceous Period.

CHAPTER 23 *Global Change in the Earth System*

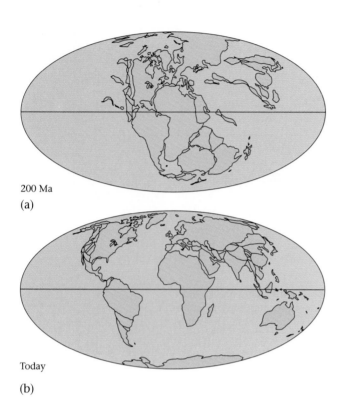

200 Ma

(a)

Today

(b)

FIGURE 23.1 A comparison of (a) a map of the Earth 200
million years ago with (b) a map of today's Earth emphasizing the
change that has resulted from continental drift.

NOTES

NOTES

NOTES

NOTES

NOTES